Aetna and the Moon

This image, depicting what is presumed to be Mt. Vesuvius and the god Bacchus (adorned in grapes), is from a wall painting from the household shrine of the House of the Centennial at Pompeii, dating from the Flavian period (68-79 CE). The image is reproduced with kind permission from the Museo Archeologico Nazionale di Napoli, inv. 112286.

Aetna and the Moon

Explaining Nature in Ancient Greece and Rome

LIBA TAUB

Oregon State University Press | CORVALLIS, OREGON

The paper in this book meets the guidelines for permanence and durability of the Committee on Production Guidelines for Book Longevity of the Council on Library Resources and the minimum requirements of the American National Standard for Permanence of Paper for Printed Library Materials z39.48-1984.

Library of Congress Cataloging-in-Publication Data
Taub, Liba Chaia, 1954-
  Aetna and the moon : explaining nature in ancient Greece and Rome / Liba Taub.
    p. cm. — (OSU Press Horning visiting scholars publication series)
  Includes bibliographical references and index.
  ISBN 978-0-87071-196-1 (alk. paper)
   1. Communication in science—Greece—History—To 1500. 2. Communication in science—Rome—History—To 1500. 3. Science, Ancient. I. Title.
   Q223.T38 2008
   501'.4--dc22
                                    2007039213

© 2008 Oregon State University Press
All rights reserved. First published in 2008 by Oregon State University Press
Printed in the United States of America

Oregon State University Press
121 The Valley Library
Corvallis OR 97331-3411
541-737-3166 • fax 541-737-3170
http://oregonstate.edu/dept/press

The OSU Press Horning Visiting Scholars
Publication Series

EDITORS *Paul L. Farber & Mary Jo Nye*

CONTENTS

A note on the spelling of Greek names and terms viii

A note on references to ancient works ix

Abbreviations x

Acknowledgements xi

Foreword, *by Mary Jo Nye* xiii

CHAPTER 1 Genres of scientific communication 1

CHAPTER 2 Scientific poetry and the limits of myth 31

CHAPTER 3 Scientific and mythic explanation in dialogue 57

Epilogue 79

A Note about Ancient 'Books' 87

Notes 91

Bibliography 112

Index 131

A NOTE ON THE SPELLING OF GREEK NAMES AND TERMS

For the most part (but not always), I have adopted a "latinized" spelling of Greek names and terms (for example, Eudoxus, rather than Eudoxos; Callippus, rather than Kallippos), to conform to general usage. Following *The Chicago Manual of Style* (15th edition), I have normally transliterated the Greek *upsilon* (υ) as 'u,' but in some cases (for example, *mythos*) as 'y.' The Greek *chi* (χ) is transliterated as 'kh' (as in *skholai*).

A NOTE ON REFERENCES TO ANCIENT WORKS

References here to Plato's *Timaeus* and Plutarch's dialogue *On the Face on the Moon* cite "Stephanus pages." This convention offers readers a uniform system of reference, regardless of which edition or translation they use, by providing in the margins of modern editions the page references to the edition of Plutarch's *Moralia* published in 1572 and of Plato's writings in 1578 by Henri Estienne (*c.* 1528/31-1598). These editions were, for more than two centuries, the standard. The name "Stephanus" comes from the latinized form of "Estienne." References to Strabo's *Geography* indicate their place in the 1620 edition of Isaac Casaubon.

ABBREVIATIONS

c.	*circa* (as in *circa* 586 BCE).
Complete Works	*The Complete Works of Aristotle*, ed. J. Barnes, 2 vols, Princeton: Princeton University Press, 1984, Bollingen Series 71.2.
DK	Diels, H. and Kranz, W. (eds) *Fragmente der Vorsokratiker*, 6th edn, 3 vols, Berlin: Weidmann, 1952.
KRS	Kirk, G.S., Raven, J.E. and Schofield, M. *The Presocratic Philosophers*, 2nd edn, Cambridge: Cambridge University Press, 1983.
LCL	*Loeb Classical Library*, London: Heinemann and Cambridge, Mass.: Harvard University Press.
LSJ	Liddell, H.G., Scott, R. and Jones, H.S. *A Greek-English Lexicon*, 9th edn, Oxford: Clarendon Press, 1940 with a supplement, 1968.

ACKNOWLEDGEMENTS

It is a very pleasant duty to acknowledge those who have contributed to the writing of this book. The essays presented here were given as public lectures when I had the pleasure of being a Horning Visiting Scholar at Oregon State University in April 2007. Prior to that, I benefited from testing out my ideas on participants on a number of occasions: a Pudding Seminar at Newnham College in 2002, a seminar on dialogue presented at the Karman Institute for Advanced Studies in the Humanities at the University of Bern in 2006, and workshops on ancient scientific texts held at Newnham College in 2006 and University College Dublin in 2007. I am grateful to the Part II students in the Department of History and Philosophy of Science who, during the autumn of 2007, read and discussed Plutarch's dialogue with me. I especially thank the audiences at my lectures in Corvallis for their helpful questions and insights.

I thank the following individuals for reading portions of the text, and offering useful suggestions and criticisms: Aude Doody, Ian Du Quesnay, Nick Jardine, Philip Hardie, Geoffrey Lloyd, Christine Salazar, David Sider, Laurence Totelin, Philip van der Eijk and Frances Willmoth. David Konstan not only read my text, offering helpful suggestions, but also generously shared his own work with me prior to publication, as did David Sedley. In addition to providing welcome suggestions for my text, Harry Hine very kindly shared his own translation of *Aetna*. Others deserve thanks for providing special help in various ways: Janet

Dudley, Marina Frasca-Spada, Ruth Horry, Tamara Hug, Natalie Kaoukji, Diana Lipton, Peter Lipton, Helen MacDonald, Joshua Nall, Maria Louise Nava, Margaret Olszewski, Torben Rees, and Jill Whitelock.

Numerous colleagues and friends provided encouragement and enthusiasm, discussing my work with me at various points, for which I am grateful. Without naming all of them, I should particularly mention: Jochen Althoff, Sabine Föllinger, Andreas Gräser, Gerd Graßhoff, Martin Kusch, Sachiko Kusukawa, Diana Lipton, Glenn Most, Ineke Sluiter, Michael Sharp and Michael Worton. A special note of thanks is due to Paul Cartledge for reading page proofs. Niall Caldwell discussed and read this work at several stages and—yet again—contributed in ways too numerous to list here.

I warmly thank Mary Elizabeth Braun and Jo Alexander at Oregon State University Press for all of their help. Finally, I thank Mary Jo Nye and Robert Nye, Thomas Hart and Mary Jones Horning Professors in the Humanities, for inviting me to give these lectures, and the series editors, Mary Jo Nye and Paul Farber, for all of their encouragement and helpful advice in producing this volume.

I dedicate this work, with gratitude, to my teachers.

LIBA TAUB

FOREWORD

The Thomas Hart and Mary Jones Horning Endowment in the Humanities at Oregon State University was established by a bequest from Benjamin Horning (1890-1991) in memory of his parents, Mary Jones and Thomas Hart Horning, who were members of pioneering families of Benton County and Corvallis, Oregon. Benjamin Horning was a 1914 alumnus of OSU who completed a medical degree at Harvard University and a degree in public health at the Johns Hopkins University. His professional career included service in public health in Connecticut, as well as a position as medical director for the W. K. Kellogg Foundation. Dr. Horning wanted his bequest at Oregon State University to expand education in the humanities and to build a bridge between the arts and the sciences.

The first Thomas Hart and Mary Jones Horning Professors in the Humanities, Mary Jo Nye and Robert A. Nye, were selected in 1994, with academic appointments in the Department of History, chaired by Paul Farber. Since then, the endowment has supported an annual lecture series, individual lectures, conferences, symposia, and colloquia, as well as teaching, research, and program and collections development. The Thomas Hart and Mary Jones Horning Visiting Scholar in the Humanities program was inaugurated in 2006 when Ken Alder, Professor of History and Milton H. Wilson Professor in the Humanities at Northwestern University, spent a week's residence on campus and presented a lecture and colloquium, while also meeting with faculty and students.

Each of the volumes in the OSU Press Horning Visiting Scholars Publication Series, under the direction of the Press's Acquisitions Editor, Mary Elizabeth Braun, is the product of a series of three public lectures given by a Horning Visiting Scholar during a one-week residence on the OSU campus. This first volume of essays was written by Liba Taub, Director and Curator of the Whipple Museum of the History of Science, Reader in History and Philosophy of Science, and a Fellow of Newnham College at the University of Cambridge. The original public lectures that form the core of *Aetna and the Moon: Explaining Nature in Ancient Greece and Rome* were given during April 10-13, 2007.

Liba Taub's research and scholarship focus on the history of scientific instruments and the history of early science, particularly ancient Greek and Roman astronomy, physics, and meteorology. Her earlier publications include *Ptolemy's Universe: The Natural Philosophical and Ethical Foundations of Ptolemy's Astronomy* (1993) and *Ancient Meteorology* (2003). She is the editor, with Frances Willmoth, of *The Whipple Museum of the History of Science: Instruments and Interpretations, to Celebrate the Sixtieth Anniversary of R. S. Whipple's Gift to the University of Cambridge* (2006) and, with Alexander Jones, of the volume *Ancient Science*, which will appear in *The Cambridge History of Science* series of eight volumes.

Aetna and the Moon elegantly and insightfully bridges the arts and the sciences in its narrative and analysis of the different genres of communication used in the ancient Graeco-Roman world for writing about natural phenomena, including a poem about the volcano Etna and a dialogue on the face of the moon. Liba Taub gives a clear and compelling description of the fluid and permeable boundaries between oral and written discourse in the ancient world and the complementarity of logical explanations of natural phenomena with mythological stories of the activities of gods and goddesses. In her specific examination of the genres of the poem and the dialogue, she convincingly shows how the so-called "Greek miracle" of scientific thinking coexisted with didactic and teleological explanations of the natural world that had their origins in poetry and myth.

MARY JO NYE

CHAPTER 1

Genres of Communicating Science

On March 23, 1989, at a press conference held at the University of Utah in Salt Lake City, two electrochemists, Stanley Pons and Martin Fleischmann, announced that they had achieved controlled nuclear fusion at room temperature with a "simple, table-top apparatus." Their announcement immediately captured attention worldwide: it simultaneously challenged the conventionally held view of fusion, a process which was thought to require extremely high temperatures or pressures, and held the promise of solving the world's energy problems.

But a long scientific controversy ensued; Pons and Fleischmann were largely discredited, and the episode is studied now by scientists, historians, journalists, and others as an example of "bad science." One immediate criticism directed against Pons and Fleischmann pointed to their style of communicating science: they had chosen a press conference, rather than a peer-reviewed scientific journal, to make their announcement. Over time, however, criticisms focused on the failure of reproducibility of the results they had claimed. In the end, the rejection of so-called cold fusion was a denunciation of both their scientific practice and content.[1]

Their choice of medium to communicate the claims regarding cold fusion, and the questioning by the scientific community and others of a press conference as a suitable vehicle for announcing scientific results (without also offering a detailed, technical account in a professional scientific journal), point to an interesting aspect of science in the late twentieth and early twenty-first centuries: a whole new range of media for communicating scientific ideas, practices, and results is now in use, including press accounts, telephone calls, faxes, emails, Web sites, blogs, and podcasts. Alongside these newer forms of scientific communication are the more traditional modern formats, especially journal articles, professional conferences, and government reports. Historians have suggested that in the cold fusion controversy distinct frames of meaning shaped different stories. Technical issues were discussed in the specialist professional scientific literature, while other issues, including some rather soap-opera-like aspects, complete with sensationalism, dominated media coverage.[2]

Scientists in the twentieth century were not the first to be able to make choices about how to communicate their work and ideas. The choices of ancient Greek and Roman authors writing about nature and scientific practice will be the subject of discussion in this volume. In this first chapter, we will look at some of the diversity of formats used by those authors to communicate their ideas about nature, and the study of nature. My position is that, while we can recognize a range of forms of discourse, we cannot make generalizations about the choice of formats, or genres, without looking at individual texts. This is what I will do in some detail in the following two chapters. In the second, I will focus on poems that not only treated the world, and what we see, as liable to natural explanations but also, nevertheless, incorporated some elements of myth. The *Aetna* poem will serve as my central text, but Lucretius' *On the Nature of the Universe* will also feature prominently. In the third chapter, I will consider Plutarch's dialogue *On the Face on the Moon* as an example of a deliberate authorial choice to juxtapose scientific explanation and myth; Plato's *Timaeus* looms large in the background as another example of such a juxtaposition, and will also be discussed.

First, let me say a little about the study of ancient Greco-Roman science more generally, before proceeding in this chapter broadly to discuss genres of oral and written scientific communication in these cultures, including poetry and dialogue.

George Sarton (1884-1956), one of the most influential historians of science in the twentieth century, noted that "one often speaks of the Greek miracle," and he explained that this phrase is "the simplest way of expressing one's wonder at the Greek achievements and one's inability to account for them." For Sarton, "the first and greatest gift of that age was a long epic in the Greek language, the *Iliad*"; the Homeric poems, the *Iliad* and the *Odyssey*, are now usually dated to the second half of the eighth century BCE.[3]

The phrase "the Greek miracle" has been variously applied to such things as the development of Athens as a political center (fifth century BCE), to ancient artistic achievement (fifth century BCE), and to ancient Greek philosophy (sixth to fourth centuries BCE).[4] In the latter case, the phrase has sometimes been used to suggest a miraculous shift from the use of mythology as a form of explanation to that of philosophy, and rational, even scientific and mathematical, explanation.[5] This idea, that the emergence of philosophy depended on emphasizing *logos* (reason) and rejecting *mythos* (myth), is not a modern invention; rather, it was first put forward by some ancient Greek philosophers themselves.[6]

However, understanding what constitutes *mythos* and *logos* is complicated. While these terms are normally translated as "myth" and "reason," some scholars have described a plurality of types of *logos*—including the rationality or logic of myth, as well as the rationality of mathematics.[7] The terms *logos* and *mythos* can, in addition, both be used to refer to a "story." *Logos* usually refers to a "story" that tries to give a plausible account of something; *mythos* may not be restricted in this way to what is possible, reasonable, or even likely. But the precise meaning—and possible ambiguities—of particular occurrences of these terms are liable to variant interpretation, and somewhat context-dependent.[8]

Furthermore, the word *mythos* should not always be read as carrying the meaning implied by the English word "myth"; it may simply mean "story," with the specific context sometimes—but not always—suggesting "fiction." Claude Calame has argued that in Classical Greece the word *mythos* did not reflect the notions conveyed by the modern category of myth; in his view the modern sense was not a category indigenous to that culture.[9] For the ancient Greeks many *mythoi* referred to stories or narratives giving an account of what they regarded as their own ancient history.[10] As Calame noted, "it is essential . . . to emphasize that the stories which the Greeks often call *muthoi* come under the heading of *logos*, 'discourse'."[11] While the term *logos* could carry meanings of "argument" and "demonstration" (amongst others, including "reason" and "calculation"), *mythos* might also be used in argument and demonstration. *Mythos* was not always described in opposition to *logos*; they were recognized forms of discourse that could, on occasion, both be invoked, and each could lay claim to the truth.[12] A distinctive feature of myth may be its "provocative ambiguity."[13] But, as Calame pointed out, "the narratives which *we* assign to the instrumental category of myth . . . depend on a symbolic process from which scientific discourse does not escape either."[14] It could be argued that some of the ancient Greeks used myth in innovative ways, to present philosophical—even "scientific"—ideas; here we could include Pherecydes of Syros (flourished 544 BCE), Parmenides (early to mid-fifth century BCE), Empedocles (*c.* 492-432 BCE), and Plato (*c.* 429-347 BCE). Not only was a movement from *mythos* to *logos* not always discernible, an opposition between *mythos* and *logos* was not always clear-cut.

Nevertheless, some ancient philosophers—and some modern scholars—chose to portray sharp distinctions between these two forms of discourse, and to argue that a shift occurred from one to the other, reflecting a marked change in thinking and understanding amongst the ancients. However, while the rejection of traditional myth was, in part, a way of presenting and validating their own work, the ancient philosophers' discourse was not entirely divorced from earlier forms of expression and explanation. Glenn Most has argued that the heritage of the earliest written Greek poetry—the poems ascribed to Homer and Hesiod—was "a decisive factor in defining the parameters of the communicative situation of early Greek philosophy." These poems served as fundamental cultural texts, available to their audiences through recitation as well as reading. Because of the deep and widespread valuing of the earliest Greek poems, it should not be surprising that communications of the early Greek philosophers "inevitably bear a striking affinity to the most prominent features of Homer and Hesiod." In some cases, early Greek philosophers deliberately chose the textual strategies of the earliest Greek poets (including, for example the use of poetic meter) to construct the texts that conveyed their ideas.[15] (The Homeric poems are usually dated to the eighth century BCE; most scholars judge Hesiod's to be later. However, from antiquity, there has been some controversy regarding the dating of the epic poems.)[16]

Sarton himself, who glossed his own views, hinted that ancient philosophy, science, and mathematics did not develop in a vacuum devoid of poetry and myth, noting that "we have used the word 'miracle' when we spoke of the sudden emergence of masterpieces like the *Iliad* and the *Odyssey*." He suggested that "the emergence and development of Greek science in the space of three centuries is not easier to explain, and hence we might use the word miracle again to express our admiration and our puzzlement."[17] Sarton regarded the great Greek epics as well as Greek science as "miracles"; poetry and science are both equally worthy of our admiration and attention.

The relationship between poetry and science is an underlying theme here: not only are ancient poetical and scientific works worthy of our attention but, when we focus that attention, there is at times a tension between science and poetry that is worth examining. However, it is not always a straightforward matter to label a work as either "science" or as "poetry"; in some cases both words will apply to the same text and, as we will see, that was clearly the intention of some authors. Poetry and science were not always in opposition in ancient texts explaining the natural world. In what follows, I offer an examination of some ancient Greek and Roman scientific texts that provide evidence to counter the suggestion that there was, in this period, a wholesale rejection of myth as a legitimate explanatory approach in favor of the adoption solely of rational and mathematical explanation. I argue that some ancient Greek and Roman scientific texts relied on and incorporated poetic forms in important ways; poets themselves were recognized as scientific authorities. More generally, I argue that our understanding of Greco-Roman science is enhanced by paying attention to the genres and styles of communication chosen by Greek and Roman authors writing about nature.

Some scholars have suggested that science developed in the ancient world as a rejection of the mythologies conveyed by the poets Hesiod and Homer. However, the status of myth, and indeed the definition of myth, was not unanimously or homogeneously agreed amongst the ancient Greeks, as the evidence of ancient authors attests; furthermore, the definition of myth is not universally agreed amongst scholars today.[18]

It is important to recognize that many early Greek philosophers did not reject poetry; in fact, some of the most influential early philosophers, including Parmenides and Empedocles, chose poetry as their medium of communication. This choice of poetry by philosophers—lovers of wisdom—may seem odd to modern readers, who are more accustomed to encountering scientific texts written in other formats, including such modern genres of scientific communication as the abstract and the poster session. Ancient Greek and Roman authors also had choices to make when setting pen to papyrus: a range of forms for communicating scientific ideas and practices was chosen by ancient authors, some of whom were innovative, creating new genres.

Before we continue, I need to clarify a few of my terms and assumptions. I define "science" as the attempt to understand and explain natural phenomena.[19] There was no one term, either in ancient Greek or in Latin, that carried the meaning of the modern English word, with its primary reference being to knowledge of the natural world.[20] The Greek word *philosophia* means "love of wisdom"; "natural philosophy" can be understood as "love of wisdom about nature." While amongst ancient authors concerned with such matters the meaning of *epistēmē* ("knowledge") could be variously understood and interpreted, Aristotle's view that *epistēmē* depended on the knowledge of causes was highly influential, not only in the ancient period, but also in later periods. According to Aristotle, *epistēmē* comprises three types of knowledge: the practical, productive, and theoretical. Theoretical *epistēmē*, in turn, can also be divided into three areas: mathematics, physics, and theology (or metaphysics).[21] Some types of *epistēmē*, as defined by Aristotle, have resonances with what is understood as "science" in later periods.

With this in mind, I will sometimes use the term "natural philosophy" to refer to efforts to explain natural phenomena. And because the ancient Greek philosophers used the word *physis* to refer to nature, and *physikoi* to refer to those who studied nature, I will sometimes use the word "physics" as the equivalent of the phrase "natural philosophy." The term "Greco-Roman science" is meant to signal that scientific work was undertaken across cultural and linguistic communities in the ancient Mediterranean world; many individuals, for example Marcus Tullius Cicero (106-43 BCE), had a working knowledge of both Greek and Latin. In many instances, scientific activities pursued by those writing in Latin were closely related to work undertaken by Greek authors. This is not to say that some Romans, who had their own interests, agendas, and approaches, did not on some occasions deliberately contrast what they regarded as "Greek" ideas with Roman values.[22] Nevertheless, Latin authors were often very self-conscious about their Greek predecessors and contemporaries.

Defining "genre"

A good working definition of the term "genre" is that offered by David Duff: "a recurring type or category of text, as defined by structural, thematic, and/or functional criteria." As Duff has noted, the term is increasingly used to classify nonliterary and nonwritten texts.[23]

In 1974 Tzvetan Todorov began a landmark article on "Literary Genres" by remarking that "the problem of genres has been discussed so often, and is today so complex, that when dealing with them one must choose between writing a book and presenting a general theoretical overview." He chose the latter solution, and argued that a genre is always part of a system, "a certain horizon of expectation, i.e., a set of pre-existing rules which orient the reader's understanding and allow him to receive and to appreciate the text"; genres "can only be defined by their mutual relations" within the genre system. Todorov was particularly sensitive to historical issues, and emphasized that a genre must be redefined in each historical period, "in accordance with the other contemporary literary genres."[24]

Two classicists interested in these topics, Gian Biagio Conte and Glenn Most, have defined "genre" as referring to "a grouping of texts related within the system of literature by their sharing recognizably functionalized features of form and content"; form and content contribute to function. They have emphasized that genre is "not only a descriptive grid devised by philological research, but also a system of literary projection inscribed within the texts, serving to communicate certain expectations to readers and to guide their understanding."[25] This view of genre points to the choices made by authors, who have a number of options open to them as they craft their texts, and recognizes also that readers have expectations that are shaped and served in different ways by different styles.

In considering ancient views on genre, Conte and Most note that the "theory of genre as such is quite lacking in antiquity." There are few theoretical discussions of specific literary genres; Conte and Most have contended that the ancients "are more interested in classifying existing works than in understanding the mechanisms of literary production and reception and are directed to the needs of the school and the library, not to the [literary] critic's."[26] Those ancient Greek authors who did write on genre were not particularly concerned with prose; poetry, drama, and rhetoric were more gripping topics, and prose may have been discussed under the rubric of rhetoric, which encompassed speeches as well as historiography.

Here, I am not attempting to develop and offer a theory of genre in general, or even of genres specific to texts that communicate scientific and mathematical ideas. However, following Duff's suggestions regarding structure, theme, and function, and Conte and Most's emphasis on the functionalized features of form and content, it seems reasonable to begin a consideration of the genres of ancient scientific texts by looking at form, content, and function to help distinguish between different types of texts. In speaking about different genres, the taxonomy I offer is arrived at by empirical means, considering particularly form and function. I concentrate on texts whose content is, broadly speaking, "scientific"—natural-philosophical and/or mathematical. However, Greek and Roman texts about natural philosophy and mathematics have different forms and various functions; it is these various forms and functions that I will explore.

Often when we approach the study of Greco-Roman science, we focus on the ideas of the natural philosophers and mathematicians. Traditionally, the study of ancient Greco-Roman science has been regarded as a part of intellectual history; until very recently, most historians of science saw themselves as producing a history of ideas. Because so much of the evidence that survives for antiquity is text-based, a reasonable approach to attempting to understand the nature of the scientific work is to focus on those texts, and in studying Greco-Roman science, the tendency has been to concentrate on the ideas conveyed in texts. While this emphasis is understandable, the focus on ideas sometimes obscures the mode of communication chosen by the author.[27] The inclination of historians of science to concentrate on ideas may eclipse our recognition, for example, that Lucretius chose, deliberately, to explain the natural world using poetry. However, by using poetry as a medium of communication, Lucretius sought to make his subject more appealing to his readers, drawing an analogy to the practice of coating the rims of cups with honey, in order to persuade children to take their medicine.[28]

Yet even though we know that Lucretius' *On the Nature of the Universe* was widely read in antiquity, and continues today to be one of the most popular pieces of Latin literature, historians of philosophy and science tend to concentrate on his Epicureanism, and his insistence on a materialist explanation of the natural world, ignoring the poetry of his "honeyed cup." Often, when his ideas are discussed and even his text is quoted, it is impossible to tell that he originally wrote in hexameter verse.[29]

I would like to ask a very general question: To what extent does the text as a whole matter, rather than "just" the ideas and concepts contained and conveyed in it? Is it only the ideas that are of interest to historians of science and mathematics, or are there other features of the text that must be considered and studied if those ideas, and the ways in which they are communicated, are to be fully understood? This is a big question, and I know I will not be able to give an adequate answer to it here. Nevertheless, I think it ought to play a part in our thinking when we read scientific and mathematical work from any period. For example, in the early modern period new forms of scientific communication were developed, notably the articles published in the newly created medium of scientific journals. These new journals affected the content of other publications as well. For example, astronomers circulated requests for observational information in the *Philosophical Transactions of the Royal Society* and then announced the results in other publications.[30] While I do not believe that it suffices to say that the "medium is the message,"[31] I also do not think that we can afford to look only for the message, while ignoring the medium. As an historian of science, I am convinced that it is worth exploring the multiplicity of formats, or genres, used to communicate scientific and mathematical ideas and methods, and not only by Greek and Roman authors.

The genre chosen by the authors to convey their ideas can tell us things about them and their intentions and expectations, about the intended audience, about the social and cultural contexts.[32] In some cases, the concentration on intellectual history and history of ideas has legitimately led us to be content-focused, but ignorance of the "vehicle" that conveys those ideas, and the totality of the work (including the genre) in which the ideas are contained, can put us in danger of being ahistorical, and of misunderstanding or distorting the message. However, there are also risks involved in focusing too narrowly on genre and related issues; such an approach would be ahistorical as well, and may result in occluding the ideas conveyed in those works being studied.

Cognizant of the inherent dangers, nevertheless I will discuss some of the different genres I have identified so far as being relevant for scientific communication in the Greco-Roman world. I will restrict myself to genres of written communication; while, certainly, oral communication about scientific ideas and practice did occur, our sources for such communication are themselves written texts, as in the case of lectures preserved as documents.

Ten genres of written scientific discourse are listed here, in no particular order, just to give an idea of the sort of variety that exists: poetry, dialogue, lecture, treatise, problem text, letter, teaching text, encyclopedia, biography, and commentary. I need to emphasize that this is not a complete list of all genres or formats used for communicating scientific ideas by ancient Greek and Roman authors; other examples might include handbook, recipe, case study, and more. In fact, while I have been working on this project the number and types of genres I have looked at have been moving targets. But I think that this list indicates the range and variety of formats employed by the ancient authors. I recognize that, while some of these genres are very familiar to modern readers, others may not be.[33]

Thinking about this list of genres, I must also emphasize that these labels cannot be taken always to represent strict divisions between formats; some dialogues, for example, Plato's *Timaeus*, which we will encounter in the final chapter, look almost like lectures. Furthermore, some texts may contain elements of a number of genres, and there are some overlapping categories. So, for example, some teaching texts are written as poems; in considering prose writings, there seem to be various types, but it is sometimes difficult to know how to divide them.

What is "scientific"?

I will focus on authors who sought to explain nature (*physis*) and the natural world, many of whom are usually considered to have been philosophers. Here I include mathematical writers, since many in antiquity regarded mathematics, like philosophy, as a branch of theoretical knowledge.[34] However, it is also important to remember that communication of scientific ideas was not confined to those we might today think of as primarily concerned with science. For example, in his *History*, Herodotus (fifth century BCE), while focusing on the Persian Wars, conveys much information regarding his own and others' ideas about various natural phenomena.[35]

I recognize that from the modern point of view what is labeled as "scientific" in the writings of the ancients is often a matter of taste. There are those scholars, for example, who claim that the ideas of the Presocratic philosophers were not scientific, while there are numerous others who would counter that they represent the beginning of the scientific enterprise.[36] I am not going to run through the arguments behind these positions here. My concern is to emphasize the range of styles and formats of texts by which scientific and mathematical ideas and methods were communicated in Greek and Latin, and to show that the forms of ancient discourse (both spoken and written) on so-called scientific topics may look very different from the modern conception of scientific discourse—but then again, perhaps not entirely different; there are often modern parallels to these genres.

Talking and writing: forms of oral and written communication

Just as today only a portion of scientific communication occurs through written forms, this was also the case in ancient Greece and Rome. For this reason, the first issue that we have to confront is the oral nature of much of Greek and Roman culture.

Orality, particularly in Greek culture, is a whole topic in itself. Since the 1920s and '30s, with the publication of work by Milman Parry and his student Albert Lord, there has been a great deal of discussion regarding the composition, performance, and improvisation of the Homeric poems. The Parry-Lord thesis emphasizing the orality of Homeric poetry was highly influential, leading some historians (notably Eric A. Havelock) to emphasize the preponderant significance of oral communication within Greek society down to the fourth century BCE. Homeric poetry is often regarded as the archetypal oral poetry. The Parry-Lord thesis considers not just the work of the poet himself, but also, importantly, the role of the audience in helping to shape the poem and the performance. This emphasis on the audience has parallels that are important in our consideration of ancient scientific discourse.[37]

Some of the discussion of natural phenomena took place literally as an oral discussion; what we have in some cases are written reports of what people said, rather than what they wrote. Thales (flourished 586 BCE), the person traditionally regarded as the first ancient Greek philosopher, who also philosophized about nature, is a good example of this. The Greek words for "being a follower of" and "listening to" are related. We are told that Anaximander (died soon after 547 BCE) was a student and "follower" (*akoustēs*) of Thales. Anaximenes (flourished 546-525 BCE) "heard" (*ēkousen*) Anaximander, but it is not clear that "heard" should be understood literally;[38] rather, the term may have simply been used to mean "was a student of." And there is the possibility that "heard" may simply mean that X was "a student/follower of" Y in a broad sense, without meaning that X literally heard Y speak or teach.[39] It is not clear whether Thales wrote anything at all; ancient authorities themselves seem skeptical about whether Thales did produce any writings. His main mode of communicating his scientific ideas (and eclipse predictions) seems to have been conversation and oral teaching.

Some of the teachings of Pythagoras were described as the "things heard" (*akousmata*; singular = *akousma*); presumably these were initially communicated orally, rather than by writing. Pythagoreanism is generally understood as having two rather distinct forms or schools after the fifth century BCE, but the distinctions between them may not have been completely clearcut. One "school," the "scientific" or philosophical form (whose advocates were the so-called *mathēmatikoi*) manifested itself in the fourth century BCE in the thinking of Philolaus and Archytas of Tarentum. The other was a religious, sectarian form, the adherents of which were known as *akousmatikoi*, those following certain oral teachings.[40] However, to attempt to dissociate the two traditions is probably a mistake, for this may give a skewed view of the range of teachings that were offered under the umbrella of Pythagoreanism. Some later authors, for example Iamblichus (*c.* 245-*c.*325 CE), preserved the "things heard" in writing, as collections of Pythagorean sayings and maxims, representing orally transmitted Pythagorean teachings.⁴¹ Pythagoreans may have been required to memorize these maxims, as a sort of catechism of their doctrines and practices. The maxims include such things as the following: Diogenes Laertius (probably first half of the third century CE) reports that "Aristotle says, in his work *On the Pythagoreans*, that Pythagoras enjoined abstention from beans either because they are like the privy parts, or because they are like the gates of Hades (for this is the only plant that has no joints)."⁴²

The oral nature of Greco-Roman culture is fundamental to understanding the spectrum of genres used for scientific discourse. Several genres embody or invoke talking, conversing, lecturing, questioning, commenting, and even singing. But in the classical period and later, some level of literacy, and even a certain degree of "bookishness," is also characteristic of ancient Greece and Rome. That the written word had various political and civic uses in antiquity is clear; the Athenian democracy relied—to some extent—on writing.⁴³ But many written texts were also read out— or performed—to an audience; Gregory Snyder has suggested that "in order to be brought to life, a text required a performer and almost always presumed the presence of an audience." Furthermore, the audience may not have been composed of passive auditors; Snyder has noted that the "audience may have partaken in the performance to a greater or lesser degree."⁴⁴ There is evidence of an increase in personal reading, for example of literary works, during the fifth and fourth centuries BCE in Greece. The sixth-century BCE tyrant, Pisistratus, was said to have established a public library at Athens, while Aristotle was credited in antiquity with having been the first to collect books (actually, papyrus rolls) for what may have been regarded as a personal library.⁴⁵

Aristotle's book collecting was related to other collecting habits he had; for instance, when he decided to work on politics, Aristotle and his students made a collection of the constitutions of the various Greek city-states.[46] A fundamental part of Aristotle's method of study was to think about what others, his predecessors and contemporaries, thought about various topics, and in many of his works, he reported and commented on the ideas and opinions of others.[47] The Greek word for opinion is *doxa*. "Doxography" is the modern term used to describe the method of collecting, writing down, and considering the opinions of others. The collection of *doxai* was used in various types of works, on different subjects, including medical as well as philosophical writings. Sometimes *doxai* were collected and reported as part of a larger project. For example, Diogenes Laertius wrote accounts of the ancient philosophers that included biographical information, as well as lists of their writings, and their opinions. Here is a sample from his "Life of Democritus": "His opinions are these. The first principles of the universe are atoms and empty space; everything else is merely thought to exist. The worlds are unlimited; they come into being and perish."[48]

The collecting and recording of the opinions of earlier thinkers may have begun in the fifth century BCE. There is some evidence that Hippias of Elis (fifth century BCE) may have recorded the views of philosophers and organized them with a view to comparing them, indicating disagreements.[49] Plato's dialogue *Theaetetus* (152e, 180e) shows traces of such organization, in contrasting the views of Heraclitus and Parmenides.[50] Doxography—the collecting and recording of opinions—can be considered in relation to another genre concerned with giving an account of the ideas and achievements of one's predecessors, namely heurematography, the writing about discoveries and inventions, and the men who were credited with them; doxography is not concerned with priority claims, which are central to heurematography.[51]

Aristotle employed doxography—the consideration of others' opinions—as part of his scientific practice; Martha Nussbaum has noted that "although we may find fault with his treatment of one or another previous thinker, he was the first Greek thinker to make engagement with the books of others a central part of his method."[52] He made specific suggestions for working with books: "we should select also from the written handbooks of argument, and should draw up sketch-lists of them upon each several kind of subject, putting them down under separate headings."[53] The lists or tables allow the classification of various reputable opinions (*endoxa*) by subject, thereby allowing ease of reference when required.

Literacy and a growing interest in books influenced the ways in which people, not only natural philosophers, worked and communicated. In the following two chapters, I will focus on two genres that especially reflect the underlying oral aspects of literate culture in the Greco-Roman world: poetry and dialogue. At this point I will simply signal the dual aspects of both. Poetry in the Greco-Roman world was often performed (recited and sung to an audience), but a culture of the book also grew up around poetry, exemplified by the growth of Homeric scholarship in the Hellenistic period. The dialogue is meant to reflect (and possibly record) a conversation; even though Plato set down his ideas on mathematics and natural philosophy in a sort of literary fiction, the dialogue,[54] he nevertheless emphasized the value of oral teaching.

In the remainder of this chapter, I will briefly outline six other genres that particularly reflect the tensions and relationships between oral and written culture. All of these were important for communicating scientific and mathematical ideas in the Greek and Roman worlds.

Lectures

The students of some of the ancient philosophers took notes and thereby recorded their oral teaching and lectures. Scholars have argued that much of what survives of Aristotle's writing is actually lectures, or notes for lectures. So, for example, the Greek title of the work known to us as the *Physics* is *Physikē akroasis*; *akroasis* may be translated as "lecture" or "hearing," rendering the title as "Lecture Course on Nature."[55] As Geoffrey Lloyd has pointed out, at some points the text is "telegraphic and takes the form of a sequence of elliptical notes," reminding us that "Aristotle's oral exposition would have elaborated the argument."[56]

Jonathan Barnes, editor of the revised version of the Oxford Translation of Aristotle's writings and the *Cambridge Companion to Aristotle*, has explained that in the case of Aristotle, "when you pick up the *Metaphysics* or the *Nicomachean Ethics*, you are not picking up a finished philosophical text." But, seeking to refute the generally held notion that what survives are lecture "notes" or "records," Barnes has quite reasonably cautioned against "the perilous supposition that Aristotle taught and worked in much the same way as a twentieth-century professor" and suggested that "it is proper to assume that you are picking up a set of papers united by a later editor; and it is proper to assume that you are reading a compilation of Aristotle's working drafts."[57] In spite of Barnes' caution, many scholars do regard Aristotle's writings as reflecting his lectures; however, it is not unusual to refer to his writings as "treatises." The boundaries between these two genres may well have been blurred because lectures, written down as notes, were edited for publication.[58]

Treatises

Prose is closely linked to literacy, and to the culture of writing texts, and it seems worth thinking about a possible genre known as "treatise," even if we are not entirely certain how such a genre was referred to in antiquity. Franz Dirlmeier, whilst considering issues related to oral and written discourse in Plato and Aristotle, suggested that the Greek word *pragmateia* has the sense of a written "work"; his understanding of what constitutes a "work" seems to coincide with our sense of "treatise." Dirlmeier argued that the appearance of the systematic, thorough written work marked a shift in styles of presenting philosophy via written texts, from the dialectical dialogues of Plato to the *pragmateiai* of Aristotle. However, as he points out, Aristotle himself appears to apply the term in referring to the title of a work only once (in *Prior Analytics* 46a30, translated by Jenkinson as "we have discussed the matter precisely in the treatise concerning dialectic").[59] Plato also used the word *pragmateia*, but it is not entirely clear when the term takes on a meaning that would distinguish a particular genre.[60] Another term that may have a similar meaning, and which was used by both Plato and Aristotle, is *suggramma*.[61] The application in antiquity of terms to refer to systematic works—what we think of as "treatises"—deserves further study. We often see ancient prose texts referred to in the scholarly literature by this modern term;[62] this may be partly motivated by the fact that the prose treatise is an important form of modern scientific communication, and one with which modern readers are most familiar.[63]

Lloyd has pointed out that "we are used to speaking of the extant Aristotelian corpus as a set of philosophical and scientific treatises";[64] but at least some of these may have begun life as lecture notes, made by the lecturer or by students. Other ancient works dealing with scientific subjects, similarly described as treatises, may also have begun as lectures; one such example is Cleomedes' work known as *The Heavens*. Cleomedes appears to have been a professional teacher; that *The Heavens* served a pedagogical purpose is indicated by the use of elementary argumentation and the frequent explication of terminology. At several points Cleomedes' language—which refers to "lecture courses" (*skholai*)—suggests that the work probably had its origin as a series of lectures.[65]

Here we have another indication of the tension between oral and written forms of discourse, and of the fluid relationships between certain genres. In attempting to come to grips with the origin of the individual texts in the Aristotelian corpus, Nussbaum—in contrast with Barnes—believes that "the most plausible view is that the extant treatises are written lectures." She thinks that "the exact wording of most of the material is Aristotle's, [but] we cannot rely on the order of books within a treatise as Aristotelian, or even the grouping of distinct books into a single treatise. All titles and many introductory and concluding sentences are likely to be the work of later editors."[66] Here, Nussbaum—with Barnes—indicates the importance of the editorial hand.

Nussbaum, like Lloyd and others, uses the term "treatise" to refer to the works of Aristotle, even while believing that they had their origin in lecture notes. This is one of the reasons I feel more confident than I might otherwise in referring, perhaps somewhat anachronistically, to a "genre" of ancient scientific discourse as "treatise." "Treatise" is a modern term; the English word refers to a written work dealing formally and systematically with a subject. (The origin is Middle English *tretis*, from Old French *traitier*, and the Latin *tractāre* meaning "handle" or "treat.") Philip van der Eijk has pointed out that the "treatise" of the Greco-Roman world is a "less well defined species of text" sometimes referred to by this modern term; its style is usually considered to be less elaborate and its formal structure does not fit in with categories of prose recognized in antiquity such as dialogues, letters, handbooks (*tekhnai*), introductions (*eisagōgai*), and commentaries.[67]

Leaving aside the question of the labels used to refer to these works, a wide variety of prose texts were written on a broad range of subjects by various authors; many of these were rather technical. Prose, as a means of literary expression, was developed in Greece (as in other places) long after poetry. Some scholars have suggested that verse forms were employed for teaching even before writing developed, because they helped the person reciting, as well as those listening, to remember what was being conveyed. Later, written didactic poetry—which would include philosophical poems—recalled preliterate verse.[68]

When we look at works referred to today as "treatises," such as those ascribed to Aristotle, and Cleomedes' *The Heavens*, we may be looking at texts that mark a shift from the orality of poetry and of lectures to an emphasis on the written word, which would have been read (sometimes aloud) and, in some cases, reproduced (either by dictation or by direct copying). The different stages of a process through which lectures become a treatise are not always clearly demarcated.

Treatises and poetry were both the subjects of another genre, the commentary, in antiquity. So, for example, commentaries were produced on the Homeric poems and on Aristotle's works; commentaries were produced on other types of texts as well, including Plato's dialogues, the Hippocratic writings, and mathematical as well as astronomical works. The treatises of Aristotle especially attracted commentators, perhaps partly because of the nature of the corpus, which is difficult to follow in places; as suggested earlier, such difficulties may have been partly due to the oral nature of the original discourse upon which the text was based. This shift from oral discourse to written texts of various formats was made possible not only by the culture of writing and reading, but also through the practice of accumulating books and creating libraries, such as that formed by Aristotle and passed on to his successors.[69] Much later, during the third century CE, there is evidence that libraries played an important role in philosophical life. In his biography of Plotinus (205-269/70 CE), Porphyry (234-*c.* 305 CE) indicates that numerous commentaries were available: "In the meetings of the school he used to have the commentaries read, perhaps of Severus, perhaps of Cronius or Numenius or Gaius or Atticus, and among the Peripatetics of Aspasius, Alexander, Adrastus, and others that were available." Not only does this list of available commentaries suggest a well-stocked library, Porphyry also makes it clear that reading of the texts was done aloud, and as a group activity.⁷⁰

Problem texts (problēmata)

Yet another genre that reflects the interactions between oral and written culture is that of the collections of "problems" or "questions and answers" (sometimes referred to as *problēmata* texts). In these texts, problems and questions are posed and answered, which may to some extent be related to the dialogic form, and hint at oral exchange, while also sharing something of the compiling of the written lists associated with doxography, in which opinions are collected and recorded. In the case of both problem texts and doxography, the form of the genre was closely related to and in part determined by the philosophical methods being advocated.

The style of philosophizing that originates in a difficulty, question, or puzzle being raised for discussion is sometimes called "aporetic" by philosophers (Greek *aporia* = difficulty, question, or puzzle). According to Aristotle (*Metaphysics* 982b), philosophy, as well as myth, is a response to wonder (*thauma*), which arises from a difficulty (*aporia*).⁷¹ He contends that:

> it is owing to their wonder that men both now begin and at first began to philosophize; they wondered originally at the obvious difficulties, then advanced little by little and stated difficulties about the greater matters, e.g. about the phenomena of the moon and those of the sun and the stars, and about the genesis of the universe.⁷²

He explains that:

> for those who wish to get clear of difficulties [for example, philosophers] it is advantageous to state the difficulties well; for the subsequent free play of thought implies the solution of the previous difficulties, and it is not possible to untie a knot which one does not know. . . . Therefore one

should have surveyed all the difficulties beforehand, both for the reasons we have stated and because people who inquire without first stating the difficulties are like those who do not know where they have to go.[73]

Aristotle himself, while he was at Plato's Academy, is understood to have compiled notes on various "difficulties" that intrigued him; this collection of problems was available to members of Aristotle's Lyceum.[74] Over time, a number of Peripatetic philosophers added to the collection. While the text known as the *Problems* in the Aristotelian corpus has the stamp of his school, the work was apparently compiled over a period of time and may not have reached its present form before the fifth century CE; in others words, it may not be the work of one individual, but many.[75]

Question-and-answer texts follow a basic pattern in which a question is posed and an answer is provided. The answers may range from rather brief (a few lines) to somewhat lengthy (the equivalent of several pages). Questions are not necessarily related to one another, although in some cases questions on similar topics are grouped together.[76] Collections of questions focusing on nature include the Pseudo-Aristotelian *Problems* and Seneca's *Natural Questions* (or *Questions about Nature*).

The Pseudo-Aristotelian *Problems* is composed of thirty-eight books, covering a wide range of subjects, from problems connected with medicine (Book 1) to those concerned with mathematical theory (Book 15), and questions about shrubs and plants (Book 20); some of the observations in the *Problems* are found also in Aristotle's works.[77] Several of the books are specifically concerned with problems related to meteorological phenomena: Book 25 deals with air and Book 26 with wind.

The problems are posed in a particular fashion; many begin with the question: "Why?"[78] Furthermore, the answers offered are generally in the form of a question: "Is it because . . .?"[79] Here is question 9 from Book 25, concerned with air:

> Why is it that, though air is denser than light, it can pass through solids? Is it because light travels in a straight line only, and so the sight cannot see through porous substances like pumice-stone, in which the pores are irregular, whereas they are not so in glass? The air, on the other hand, is not obstructed, because it does not travel in a straight line through anything through which it passes.[80]

This is the sort of question we can imagine being answered by one of the *physikoi*. Book 15 deals with questions concerning mathematical theory, for example, question 10:

> Why are the shadows thrown by the moon longer than those thrown by the sun, though both are thrown by the same perpendicular object? Is it because the sun is higher than the moon, and so the ray from the higher point must fall within that from the lower point? Let AD be the gnomon [shadow caster], B the moon, and C the sun. The ray from the moon is BF, so that the shadow will be DF; but the ray from the sun is CE, and its shadow therefore will necessarily be less, viz. DE.[81]

Here, a question arising from observation—the length of shadows—is answered by means of a geometrical demonstration.[82] One of the reasons I chose this question as an example is because it refers to another piece of writing—in this case a drawing or diagram, of which, to my knowledge, no contemporary version survives. However, there are similar references to such visual aids in works by Aristotle, for example the *Meteorology*; such diagrams may have been included in a text, or displayed to an audience during a lecture.[83] Once again, a genre of written text—the *problēmata*—was also linked to oral discourse and presentation.

Some forms of question-and-answer texts display links to other written genres. The *Natural Questions* produced by Seneca (who lived between 4 BCE and 1 CE - 65 CE) deals mainly with natural phenomena and sometimes conveys ethical messages; this text is more elaborate, and unlike the Pseudo-Aristotelian *Problems* is not simply presented as a written list. The *Natural Questions* reads, in places, like a treatise; in some places it can be read almost as a long letter, addressed to Seneca's friend Lucilius. The *Natural Questions* is an excellent example of a text that shows affinities with several genres used to write about nature: question-and-answer text, letter, and treatise.

There is also an argument for suggesting that certain logical and mathematical texts (for example geometrical propositions and proofs) can be understood as related to question-and-answer texts; in certain geometrical texts, for example Euclid's *Elements*, problems are presented and solved. (The *Elements* is usually regarded as a compilation, perhaps collected by Euclid [between 325 and 250 BCE].) There is no distinct genre of "mathematical text" in antiquity; those authors writing about mathematics used a variety of formats, including some that look similar to question-and-answer texts; other texts are deliberately cast as letters, in some cases addressed to specific individuals, including patrons.[84]

Letters

Various sorts of letters were written and circulated in antiquity; not all letters were intended for private purposes.[85] Some, like the letters published today in newspapers, were clearly meant for a wider readership. Letters were used for various purposes, including those intended to give philosophical advice and instruction, such as the three letters of Epicurus (341-270 BCE) preserved in the "Life of Epicurus" by Diogenes Laertius. Other letters contained technical or scholarly work on mathematical, mechanical, and medical topics. A number of letters written by ancient Greek mathematicians survives, indicating that letter writing was a useful mode of communication for them; Eratosthenes of Cyrene (*c.* 285-194 BCE), who lived in Alexandria, was the recipient of letters from Archimedes (*c.* 287-212 or 211 BCE), living in Syracuse.[86]

Some of the letters are clearly communications between friends and colleagues, and have almost the flavor of a conversation; others, particularly the letters of Epicurus, were intended to be instructional, serving as brief summaries of his views for his students and followers.

One of his letters, preserved by Diogenes Laertius (Book 10, 83-85), begins as follows:

> Epicurus to Pythocles, greeting.
> In your letter to me, of which Cleon was the bearer, you continue to show me affection which I have merited by my devotion to you, and you try, not without success, to recall the considerations which make for a happy life. *To aid your memory* you ask me for a clear and concise statement respecting celestial phenomena; for what we have written on this subject elsewhere is, you tell me, hard to remember, although you have my books constantly with you. I was glad to receive your request and am full of pleasant expectations. We will then complete our writing and grant all you ask. Many others besides you will find these reasonings useful, and especially those who have but recently made acquaintance with the true story of nature and those who are attached to pursuits which go deeper than any part of ordinary education. So you will do well to take and learn them and get them up quickly along with the short epitome in my letter to Herodotus.[87]

Epicurus' *Letter to Pythocles* refers to other letters and correspondence; not only have Epicurus and Pythocles exchanged earlier letters, there is also a reference to his *Letter to Herodotus* being read and used by Pythocles as well. As Epicurus indicates to Pythocles, he expected that others would also find these letters useful; his letters were not intended as private communications, but were meant to be read by a wider audience. Epicurus' *Letter to Pythocles* and *Letter to Herodotus* served as summaries of his natural philosophy, and could function as introductory teaching texts, presented in the form of a letter.

Introductions (teaching texts) (eisagōgai)

There was clearly a market for pedagogical works in the Greco-Roman world. Certain works were intended to serve as introductions (*eisagōgai*) or teaching texts; several of these survive, including Nicomachus of Gerasa's (between 50 and 150 CE) *Introduction to Arithmetic*, which was an elementary text on mathematics, and Geminus' (*c.* 50 CE) *Introduction to the Phenomena*. Both works begin with definitions. Geminus' text offers the following (1.1-2):

> The circle of the signs is divided into 12 parts, and each of the sections is designated both by the common term "twelfth-part" and by a particular name taken from the stars that it contains and by which each sign is formed. The twelve signs are: Aries, Taurus, Gemini, Cancer, Leo, Virgo, Libra, Scorpio, Sagittarius, Capricorn, Aquarius, Pisces.[88]

Here, as in some other texts we have considered, it is relatively easy to imagine that Geminus' *Introduction* would have supplemented lectures; students may well have appreciated a written text to consult before and after the oral presentation.

Nicomachus' *Introduction to Arithmetic* also begins with definitions: "The ancients, who under the leadership of Pythagoras first made science systematic, defined philosophy as the love of wisdom."[89] Along with this work on arithmetic, Nicomachus produced an *Introduction to Harmonics*, an *Introduction to Geometry* (which has not survived), and possibly an *Introduction to Astronomy*. His *Introduction to Arithmetic* was used as a teaching text throughout later antiquity, and into the Middle Ages (in a Latin paraphrase produced by Boethius [*c.* 480-c. 524 CE]); a number of commentators, including Iamblichus (*c.* 245-c. 325 CE), Asclepius of Tralles (died *c.* 560/570 CE), and Philoponus (*c.* 490-570s CE), wrote about the work, indicating that it was the focus of further study itself.⁹⁰

Commentaries

As part of the developing literary culture of the "book," the didactic and scholarly traditions produced a variety of handbooks, epitomes, and commentaries; the works of Aristotle and mathematical texts (including Nicomachus') were often the subject of such treatments. Aristotle (*Topics* 1.14) makes it clear that he is working within a culture of the "book," a culture of both reading and writing. He instructs that "in the margin, too, one should indicate also the opinions of individual thinkers, e.g., that Empedocles said that the elements of bodies were four," thereby indicating the desirability of active involvement by the reader with the text.⁹¹

Detailed scholarly exegesis of the Homeric poems was underway by the third century BCE, and eventually philosophical and mathematical texts (as well as medical works) were also the focus of some very detailed attention. While commentaries on various types of texts were important from the third century BCE, the commentary was a particularly significant genre for scientific writing in the later period. The sixth century CE was an especially important period for commentaries, and Alexandria was a significant site for this tradition. The commentary continued to be a key genre for the medieval period.

Typically, a passage from the ancient source is quoted, and then a comment appended, which may be of any length, from one sentence to several pages. Additionally, the commentator may refer to other works, by the author of the target text, or other writers. So, for example, in his commentary on Aristotle's *Physics* (193b23), Simplicius (sixth century CE) comments on the following passage (his longer comment is only briefly quoted here): "We must next consider in what way the mathematician differs from the physicist." Simplicius notes that Aristotle "quite justifiably wants to show the difference between the physicist and the mathematician, since they appear to concern themselves with the same subjects."[92] He discusses the passage at length, making references to both Aristotle's *On the Heavens* and Plato's *Timaeus*. Even as commentaries encouraged a close engagement with particular texts, they often served as vehicles for the presentation of the commentator's own ideas.[93]

The commentary in some ways represents the culmination of the movement from oral forms of discourse to the establishment of new written traditions, which are in themselves text-focused. However, commentaries often functioned within teaching contexts, in which lectures and discussion took place; in his biography of his teacher Plotinus, Porphyry reports that "in our gatherings he would have the commentaries read out to him."[94] A number of important commentators on Aristotle's works were publicly supported teachers (or "professors"), including Alexander of Aphrodisias (flourished early third century CE), as well as John Philoponus and Olympiodorus in the sixth century CE; the public funding of these posts indicates the value that was placed on education in philosophy, including natural philosophy.[95] Alexander and Olympiodorus were appointed to teach philosophy, while Philoponus was, officially, a grammarian (or philologist).

I began this chapter with an example of a very modern form of scientific communication—the press conference. However, as we know, even in the time since that press conference in 1989 several new forms of media have been invented. Historically, styles of scientific and mathematical communication have been influenced by various societal and cultural factors; technology has also played an important role. In antiquity, the shifts from primarily oral to largely written discourse about nature were more gradual, but new media, forms, and genres were created. In what follows, I will focus on poetry and dialogue, two ancient genres that particularly play on the tension between oral and written communication.

P· VIRGILII MARONIS AETHNA INCIPIT·

Aethna mihi ruptique cauis fornacibus ignes
Et quae tam fortes uoluant incendia causae
Quid fremat imperium quid raucos torqueat estus
Carmen erit· doceer uenias mihi carminis auctor
Seu te cynthos habet· seu delos gratior illa
Seu tibi dodona potior· tecumq; fauentes
Innoua· pierio pperent a fonce sorores·
Vota per insolitum phoebo duce tutius iter
Aurea securi qui nescit saecula regis
Cum domitas nemo cererem lactaret in aruis
Mercurius malis prohiberet fluctibus herbas
Annua sed saturnae complerent horrea messes
Ipse suo flueret bacchus pede mellaq; lentis
Penderent foliis etpingui pallas oliuae
Secretos amnis ageret· cum gratia ruris
Non cessit· cuiquam melius sua tempora nosse
Ultima quis tacuit iuuenum certamina colchos·
Quis non argolico defleuit pergamon igni
Inpositam· et tristi natorum funere mentem·

This image is of the opening of *Aetna* in a tenth-century manuscript (Kk.v.34, folio 95), held in the University Library, University of Cambridge. Reproduced by permission of Cambridge University Library.

CHAPTER 2

Scientific Poetry and the Limits of Myth

In the Greco-Roman world, poetry was a particularly valued genre, not least because of the far-reaching cultural importance of the archaic epic poets, Homer and Hesiod.[1] While prose forms often dominated the technical literature, a significant number of widely read poems communicated many types of scientific information, technical data, and instructions on a range of scientific, mathematical, technological, and medical subjects.

Poetry describing or explaining nature appears in ancient texts in different forms. There are poems that have as one of their primary functions the communication of information about the natural world—that provide descriptions and explanations. I would include, for example, Lucretius' poem *De rerum natura*, the title of which is translated as either *On the Nature of Things* or *On the Nature of the Universe*, as well as Book Ten of Columella's *On Agriculture*, which mainly deals with horticulture.[2] There are also examples of poems dealing with mathematical topics.[3] And there are the quotations of poetic passages inserted into texts by prose authors, included for a variety of reasons, in some cases to refute the ideas of the poet, in others to cite the poet as an authority.[4] In both of these cases, prose authors engage with poetry as a source of knowledge. But there remains a question: Why would an author choose poetry as the genre to explain the natural world?[5]

In attempting to answer this question, I will focus on a particular Latin poem about the volcano Etna. The author of this poem is not known, nor is the dating secure. I have chosen this poem because it is an excellent example of a scientific text. I will argue that the *Aetna* poet had several aims:

(1) to explain a natural phenomenon—in this case a particular volcano, Etna;

(2) to encourage others to participate and engage in the study of natural phenomena;

(3) to offer advice and examples regarding good "scientific" practice (while also providing examples of what should be avoided).

This poem has attracted the attention primarily of classicists; few historians of science have paid much attention to it, a situation which was bemoaned by P. B. Paisley and D. R. Oldroyd. They stressed that the poem "avows an intention of maintaining purely naturalistic explanations of phenomena," while noting that it "includes a number of gestures towards popular mythology, without any sarcasm or censoriousness."[6] As will become clear in what follows, I disagree: the *Aetna* poet thinks that myths may be only lies, and that an interest in mythology deters people from more important pursuits. But, having said that, I must admit that the *Aetna* poet does, at a couple of points in the poem—namely the beginning and end—refer respectfully to the gods, and their influence on man.

In focusing on the *Aetna* poet and his concerns, my aim is to question the idea that in the Greco-Roman world poetry and mythology were rejected in favor of philosophy as a way to explain the world. In doing this, I will point to the complicated texture of some ancient Greek and Latin poems on nature, in which the use of myth and the roles of gods are not completely rejected, even as rational, natural philosophical accounts of nature are offered. It is not my goal to try to explain what the traditional gods are doing in these poems, but to point out that they are there (and probably for different reasons for different poets). In any case, the gods have not been totally excluded, nor have they disappeared.

The choice of genre makes this work particularly intriguing, because the author, who self-consciously presents himself as a poet, criticizes poetry as conveying false legends. Near the beginning of the poem (lines 29-35), the poet cautions:

> First, let none be deceived by the fictions poets tell—that Aetna is the home of a god, that the fire gushing from her swollen jaws is Vulcan's fire, and that the echo in that cavernous prison comes from his restless work. No task so paltry have the gods. To meanest crafts one may not rightly lower the stars; their sway is royal, aloft in a remote heaven; they reck not to handle the toil of artisans.[7]

He warns his readers not to believe the fictions of poets: that Etna is the work of Vulcan (in the extract above), or that the Cyclopes used Etna to forge Jupiter's thunderbolt (36-40), or that Etna played a role in the conflict between the Giants and Jupiter (41-73). Roughly eighty lines of the 646 of the poem are devoted to railing against the fictions of poets. While he acknowledges that some poets have genius (line 75: *vatibus ingenium est*) and that poets must be allowed their license (91: *debita carminibus libertas ista*), he proclaims that "truth alone is my concern" (91-92: *sed omnis in vero mihi cura*).

Before focusing on the *Aetna* poem, it will be useful to consider briefly other ancient accounts of the gods and myth, and to examine the poetic traditions within which the *Aetna* poet wrote.

The gods and myth

Some modern historians and philosophers would discount mythological accounts as being of no interest from a "scientific" standpoint; such a view has a distinguished history. Aristotle seems to have been the first to have directly contrasted poets and storytellers recounting myths about gods and heroes (*mythologoi* and *theologoi*) with the *physikoi* and *physiologoi*, philosophers who investigated the natural world.[8] But, in spite of the clear opposition proposed by Aristotle, the relationship between mythology and philosophy was in fact more complicated, as the writings of the ancient philosophers themselves demonstrate.

Myth was treated at times by some ancient philosophers, including Plato, as an acceptable form of description and explanation. Both Plato (*Cratylus* 402b; *Theaetetus* 152e, 180c-d) and Aristotle (*Metaphysics* 983b27ff.) suggested, perhaps jokingly, that Homer and Hesiod were the fathers of ancient philosophy. W. K. C. Guthrie noted that "Plato was fond of calling Homer the ancestor of certain philosophical theories because he spoke of Oceanus and Tethys, gods of water, as parents of the gods and of all creatures."[9] While such pronouncements may have been made lightly, there may also have been an element of seriousness. Within some definitions of mythology, the accounts of the traditional gods and their activities may be understood, quite reasonably, as a form of explanation. There was an ancient tradition of regarding the earliest poets, Homer and Hesiod, as intellectuals; standing at the fountainheads of tradition, they helped shape intellectual agendas. Ancient philosophers were by no means unanimous in their views of the archaic epic poets, but the variety of reactions to them indicates a need to engage with their accounts of the world.[10] One approach adopted by some ancient readers was to provide a "rationalizing" interpretation of the epic poems, through adopting an allegorical reading; as an example of this approach, the third-century CE philosopher Porphyry pointed to the suggestion by Theagenes of Rhegium (flourished *c.* 525 BCE) that in the *Iliad* (20.67) Homer spoke allegorically rather than literally, identifying gods with the hot and cold, dry and moist. Allegorical explanations erode the distance between what might be regarded as "rational/scientific" and what is "mythic."[11]

The earliest surviving Greek texts, the Homeric and Hesiodic poems, provide evidence that many phenomena were traditionally linked to the gods. The use of mythology to explain what we regard as "natural" phenomena is an important part of the fabric of Greco-Roman culture; Zeus, for example, is described as "thunder-bearer." Many of these phenomena, including storms, lightning, and thunder, are potentially dangerous and particularly frightening; traditional mythology offered explanations of meteorological events as acts or epiphanies of gods, and traditional religion provided ways to cope with the danger and fear.

Some scholars have argued that Greek natural philosophy excluded gods as agents in the natural world; however, this so-called exclusion is an oversimplification.[12] In most of the ancient philosophical schools there was an assumption of some "divine" (that is, eternal) presence in the cosmos; in some cases, the cosmos itself was thought to be divine. The nature of this "divinity" (or these "divinities") was a central concern for many ancient philosophers, and it is important to recognize that what is "divine" was not necessarily understood as being a "god." In those philosophical schools in which the existence of gods was accepted, the role of gods (and the divine) was understood in a number of different ways. The divine (or eternal) was not always represented through anthropomorphized gods; some philosophers regarded the planets as divine bodies. Nor were the gods always understood as playing an active role in the universe: the Epicureans' gods existed in a blissful state, unconcerned with the affairs of humans.[13] In the lines quoted earlier from *Aetna* (29-35), the poet proclaims that the gods are too elevated to lower themselves to be bothered with such mundane things as volcanoes.

Even though in some authors' works the links between the gods and natural phenomena were severed in favor of other, naturalistic, explanations, the epic poets were still often revered as offering relevant knowledge. The long-lived authority of Homer and Hesiod on natural phenomena is attested by the many quotations and allusions to their poems, even in technical prose treatises. For example, on the topic of stars and weather signs, some writers, including Pliny the Elder (died 79 CE), considered Hesiod to be as valid an authority as some specialist astronomers.[14]

From poetry to philosophy?

Surviving textual evidence weighs against the idea that those who were interested in explaining the natural world suddenly shifted from poetry to philosophy. One of the earliest philosophers, Parmenides of Elea (born *c.* 515 BCE), presented his ideas in a single, and extraordinarily influential, work: a poem, which survives only in fragmentary form.[15]

The rhythm of ancient poetry was defined by its meter, and relied on the arrangement of syllables, and the length of time taken for pronunciation, to determine the rhythmic pattern. The most important meter used in classical poetry is the dactylic hexameter, the meter of Homer and Virgil. This is the meter chosen by Parmenides and many of the poets who offered explanations of the natural world; by choosing this form, they associated themselves with the epic poetic tradition.[16]

Parmenides' influence on philosophy was very powerful. As Malcolm Schofield has noted, Parmenides "had an enormous influence on later Greek philosophy; his method and impact have rightly been compared to those of Descartes' *cogito*." But Schofield also cautioned that "ancient and moderns alike are agreed upon a low estimation of Parmenides' gifts as a writer. He has little facility in diction, and the struggle to force novel, difficult and highly abstract philosophical ideas into metrical form frequently results in ineradicable obscurity." However, Schofield admits that "in the less argumentative passages of the poem he achieves a kind of clumsy grandeur."[17]

Here is a small sample from the beginning of Parmenides' poem:

> The mares that carry me as far as my heart ever aspires sped me on, when they had brought and set me on the far-famed road of the gods, which bears the man who knows over all the cities. On that road was I borne, for that way the wise horses bore me, straining at the chariot, and maidens led the way.[18]

Even in a prose translation Parmenides' words do have a poetic grandeur. But why did he choose poetry to present his philosophical views? By echoing the diction and meter of the epic poems attributed to Homer and Hesiod, Parmenides was placing himself within a powerful tradition, but it is not entirely clear whether he was attempting to enlist their authority, or to subvert it.

As Parmenides' poem continues, he explains that he was directed by a goddess. The presence of the goddess here, at the beginning of his work, is significant; we will encounter other gods and goddesses mentioned at the start of other philosophical poems focusing on nature. The notion of divine inspiration, of a god or goddess functioning as a muse even for philosophers, is powerful and traditional. This divine inspiration is often invoked at the beginning of the poem, in what is called the "proem," the term used to refer to an introduction, prelude, or preface, especially at the beginning of a book and often found at the beginning of poems explaining nature. The word derives from the Greek *pro*, meaning "before," and *oimē*, meaning "song."[19]

Alexander P. D. Mourelatos, in *The Route of Parmenides*, addressed the question as to why Parmenides chose to express his views in the form of hexameter poetry, noting, "he could, after all, have written a treatise." Mourelatos argued that the question is not really relevant, for "if Parmenides is indeed using the vocabulary, structures and patterns of the poetic tradition as a model toward the development of new concepts, then the 'could' of 'he could have done all this in prose' is a modality so weak as to be meaningless." Mourelatos argued that the *logos* of "prose would not have been a live option for one whose very concept of knowing was based on an analogy with 'questing' and 'journeying'."[20]

Empedocles (*c.* 492-432 BCE) was another early philosopher who chose poetry as his medium. He is sometimes regarded as an emulator of Parmenides and also wrote in hexameter verse.[21] Aristotle, who was interested enough in the subject to write a work called *Poetics*, did not have a high opinion of Empedocles as a poet. In his view, "Homer and Empedocles have nothing in common but meter; so while it is right to call the one a poet, the other should be called a physicist, rather than a poet."[22] But other ancient readers and writers disagreed; Plutarch (born before 50 CE, died after 120) thought Empedocles a rather powerful poet. He argues that

> Empedocles is not in the habit of polishing up facts with the showiest epithets he can find for the sake of fine writing, as if he were laying on gaudy colours; instead he makes each expression reveal an essence or a power of something, as for example "mortal-enclosing earth" (of the body surrounding the soul), "cloud-gatherer" (of the air), and "rich in blood" (of the liver).[23]

These epithets are reminiscent of the Homeric poems ("wine-dark seas"). Amongst modern readers, Schofield has commented on Empedocles' poetic style, noting that he "exploited his chosen medium to express and reinforce the complex unity of his vision of the world by means of two devices in particular: a novel employment of the Homeric technique of repeated lines and half-lines, and an equally individual use of simile and metaphor."[24] Aristotle criticizes Empedocles as a poet, but did not find his ideas without value. For example, in *On the Generation of Animals,* Aristotle praises him, noting that "Empedocles puts this well in his poem, when he says: 'Thus do tall trees bear eggs: first olives . . .'; for the egg is a foetus, and the animal is produced out of part of it, while the remainder is nutrient."[25] But Aristotle's assessment of Empedocles may be an indication that—by Aristotle's time—there was some question as to whether poetry was an appropriate vehicle for philosophical ideas; this may be why he insists that Empedocles is not a poet, but a physicist. Nevertheless, poetry was clearly regarded by these authors as a perfectly legitimate medium for communicating what we might regard as scientific material.[26] There is no sign that someone suddenly said, "Now we're doing philosophy, so let's say good-bye to poetry."

An indication of the important and varied roles of poetry was provided by the Greek writer Strabo (born about 64 BCE, died after 21 CE), known for his work on geography. He explained that

> Eratosthenes contends that the aim of every poet is to entertain, not to instruct. The ancients assert, on the contrary, that poetry is a kind of elementary philosophy, which, taking us in our very boyhood, introduces us to the art of life and instructs us, with pleasure to ourselves, in character, emotions and actions. And our School [presumably, that of the Stoics] goes still further and contends that the wise man alone is a poet. That is the reason why in Greece the various states educate the young, at the very beginning of their education, by means of poetry; not for the mere sake of entertainment, of course, but for the sake of moral discipline.[27]

That poetry did not only entertain, but also served a crucial educational function, is clear; that education was the deliberate aim of certain authors is also evident from their poems. Twice in *De rerum natura*, Lucretius stated that he thought using poetry as a medium of communication might make his subject more palatable to his readers. As mentioned in Chapter 1, he drew an analogy to the way physicians coat the rims of cups with honey, to persuade children to take medicine.[28]

For the ancient Greeks and Romans, poetry could serve as a source of scientific (and even quite technical) ideas. Some poets sought to teach through their poems. But some of the authors of technical and scientific poems were more celebrated for their artistic talents than their scientific interests; Virgil, not usually regarded primarily as a "scientific" or philosophical author, wrote on "scientific" topics in the *Georgics* (including the subjects we refer to as agriculture, astronomy, and meteorology). While the content was important to the authors (and presumably the readers), so was the poetic form; these works are not just philosophy or science, but poetry. Poetic self-consciousness is often very apparent in the poems explaining the natural world, sometimes serving to establish a poetic pedigree, for example to set the poet within the tradition of Hesiod and other poets. However, Hesiod's own legacy is complicated: he was regarded as a poet and author who sought to explain the origin of the world (in the *Theogony*) and to give advice to men regarding their own lives (in the *Works and Days*).

Aratus (*c.* 315 to before 240 BCE), the poet responsible for the *Phaenomena*, was apparently assigned the task of setting the prose work of Eudoxus (*c.* 390-c. 340 BCE) to verse by his patron, Antigonus Gonatas of Macedonia;[29] his own interest in the subject is not clear, and the content of the poem was criticized by the astronomer Hipparchus (flourished second half of second century BCE). Yet Aratus' poem, along with Lucretius' *De rerum natura*, would likely be on everyone's top ten list of important works written about the natural world in antiquity. The number of Latin translations (at least four) of the *Phaenomena* attests to its popularity in antiquity, and Lucretius is today one of the most widely read authors of Latin literature. These poems, and others, including Manilius' *Astronomica* (early first century CE), are sometimes described as "didactic" poetry by (some) ancient as well as modern authors, and share specific characteristics, most notably an explicit intent to teach.³⁰ But it is important not to lose sight of the other poetic features of these works. For example, the meter of choice for didactic poetry was usually the epic hexameter, a choice that automatically signaled an affinity with other epic poets, including Homer and Hesiod.³¹

Another feature of didactic poetry is the hymning that often occurs; a hymn to a god or goddess may be featured in a proem (and elsewhere). Our English word "hymn" is a simple transliteration of the Greek, but the relation of Greek to English hymns is not at all simple. The Greek word *hymnos* can refer to a song of any kind, but especially to a song in honor of a god; in the Greco-Roman world, some individual humans were treated as deities, and so were hymned as well.³² The earliest hymns, the *Homeric Hymns* dating from before 400 BCE, are also described as preludes (*prooimia*), because originally they were sung as introductions to other hexameter compositions. Typical features of hymns include: lists of the god's powers, interests, tastes, and favorite places; accumulations of the god's epithets; and portrayals of the god in characteristic activities.³³

Lucretius' poem, *De rerum natura*,³⁴ begins with a proem, in which the goddess Venus is hymned and invoked as a muse (lines 1-2; 20-25):

> O mother of the Roman race, delight
> of men and gods, Venus most bountiful,
> . . .
> Since you and only you are nature's guide
> And nothing to the glorious shores of light
> Rises without you, nor grows sweet and lovely,
> You I desire as partner in my verses
> Which I try to fashion on the Nature of Things.³⁵

This proem has attracted a great deal of scholarly attention because here the Epicurean poet, who strenuously denies a role for the gods within our world, prays to a goddess to inspire his poem. Later in the poem (2.646-51), Lucretius further articulates his views regarding the role of gods:

> For perfect peace gods by their very nature
> Must of necessity enjoy, and immortal life,
> Far separate, far removed from our affairs.
> For free from every sorrow, every danger,
> Strong in their own powers, needing naught from us,
> They are not won by gifts nor touched by anger.[36]

This paradoxical situation—of calling on a god for help while asserting that gods have no active role in the world—has vexed historians of ancient philosophy, as well as literary scholars; there have been various attempts to explain it.[37] An allegorical reading of some of the references to Venus is possible,[38] but Lucretius himself (2.655-60) warns that gods' names can be used allegorically only if care is taken to avoid any false religious implications:

> If anyone decides to call the sea Neptune,
> And corn Ceres, and misuse the name of Bacchus
> Rather than give grape juice its proper title,
> Let us agree that he can call the earth
> Mother of the Gods, on this condition—
> That he refuses to pollute his mind
> With the foul poison of religion.[39]

The Epicureans did not argue against the existence of divinities, but asserted that gods have no active role in the universe: they are too busy being blissful, in the interstices between worlds, to be bothered with earthly concerns.[40]

So why did Lucretius begin his poem with a hymn to Venus? A concern with his literary pedigree may have influenced him: the earliest Greek didactic poem, Hesiod's *Works and Days*, began with a hymn to Zeus. But pointing to the possibility of literary imitation does not resolve the paradox of the proem. The Epicurean view of the gods is that they are peaceful and detached from worldly cares; David Sedley has suggested that Lucretius' invocation of Venus was meant to encourage in his Roman readers a belief in peaceful, detached gods, rather than angry and warring gods, like Mars. Lucretius' invocation of Venus may then be read as a moral use of myth and prayer.[41]

This reading is appealing, not least because the Epicurean project was centered on ethics, confronting questions of how to live in the world. The ethical goal of Epicurean philosophy was to live free from anxiety. One way to eliminate anxiety was to banish fear of gods, and their unpredictable interventions in our world. It was the goal of eliminating fear and anxiety that motivated Lucretius to explain Epicurean philosophy and physical theory in his poem. In the final portion of the work, Book Six, he discusses specific celestial, meteorological, and terrestrial phenomena that are sometimes regarded as terrifying wonders, and offers rational explanations for them; his goal is to eliminate the search for supernatural causes. Throughout his poem, Lucretius shows that the basis of the universe is material and natural. While human beings are in some circumstances tempted to believe that supernatural or divine beings, such as the gods, are responsible for the creation and workings of the world, such explanations, in his view, not only are unnecessary, but are wrong—for they misrepresent the nature of the gods—and are harmful.[42] Once humans understand the material workings of the natural world, there is no need to fear the supernatural intervention of gods, and *ataraxia* (freedom from anxiety) is possible. Lucretius presents these views in the "honeyed cup" of his poem.

Another group of Hellenistic philosophers, known as the Stoics, competed with the Epicureans in the philosophical marketplace. The Stoic worldview tended, by contrast, to emphasize the permeation of the cosmos by the divine.[43] The poet Marcus Manilius,[44] influenced by Stoic philosophy, composed the *Astronomica*, an astronomical and astrological poem in which he presents his view that the divine spirit pervades the entire universe and that the universe reflects the *ratio* of the divine will. The unity of the natural world, and its dependence on the divine spirit, is his underlying theme. Manilius was clearly familiar with Lucretius' *De rerum natura*; he may have intended his own poem, in part, as an attack on Lucretius' Epicurean ideas.[45] But Manilius' explanation of natural phenomena does share something with that of Lucretius, in that even potentially terrifying phenomena are part of the natural order; their causes can be understood by humans and therefore need not be feared. And, like Lucretius, Manilius chose the poem as his medium for presenting his views.

Like Lucretius, but perhaps less surprisingly, Manilius also begins his poem with an invocation of gods:

> By the magic of song to draw down from heaven god-given skills and fate's confidants, the stars, which by the operation of divine reason diversify the chequered fortunes of mankind; and to be the first to stir with these new strains the nodding leaf-capped woods of Helicon, as I bring strange lore untold by any before me: this is my aim. (1-6)

> A more fervent delight is to know thoroughly the very heart of the mighty sky, to mark how it controls the birth of all living beings . . . and to tell thereof in verse with Apollo tuning my song. (16-19)

The importance of communicating through poetry is emphasized: the first word of the *Astronomica* is *carmen* ("song" or "poem"). The poet invokes Apollo as his muse; Mercury will be called upon later. Manilius further proclaims that:

> Two altars with flame kindled upon them shine before me; at two shrines I make my prayer, beset with a twofold passion, for my song and for its theme. (20-22)[46]

We have seen even Lucretius, who rejected any active role for gods in the universe, praying to Venus; here, Manilius calls to Apollo and Mercury as his muses, while marking his commitment both to his poem (as poetry) and to its philosophical content.[47]

Aetna

As noted earlier, the identity of the author of the *Aetna* poem is unknown. While various scholars have made suggestions regarding the authorship, here the focus will not be on that question. The dating of the poem is also not secure, but many scholars argue that it must have been written prior to the devastating eruption of Vesuvius in 79 CE, as that eruption is not mentioned in the poem.[48]

There is the possibility that the subject of the poem was not driven by the poet's own interest. Just as Aratus was set his poetic task by a patron, the *Aetna* poet may have been responding to a request to set ideas in verse; this was also the case, for example, with Columella's poem on gardening (Book 10 of his *On Agriculture*). (However, we know that Columella was interested in gardens, even if writing a poem on the topic was not his own idea—or so he says.)[49] The first-century CE Latin author Seneca, who was particularly interested in meteorological phenomena, asked his friend Lucilius to include Etna in his poem about Sicily, as Virgil, Ovid, and the Augustan poet Cornelius Severus had done in poems of their own.[50] It is not clear that Lucilius was the author of the *Aetna* poem we have, but whoever that poet was may have taken up his task in response to the request of a friend or patron.

Even though we do not know exactly when he wrote, or who he was, our author here has chosen poetry as his medium, yet is very critical about the fictions that poets promote. The *Aetna* poet seeks to set himself apart from (at least some) other poets. At the beginning of his poem he condemns poets for relying on mythological gods to explain nature. This criticism reappears towards the end of the poem, as he launches into a tirade against those poets who encourage a sort of cultural tourism linked to mythology:

> Over the paths of the sea, through all that borders on ghastly ways of death, we hasten to visit the stately glories of man's achievement and temples elaborate with human wealth or to rehearse the story of antique citadels. Keenly we unearth the falsehoods told by ancient legend and we like to speed our course through every nation. (569-73)

As one example of many, he notes that
> Moreover, Greek paintings or sculptures have held us entranced.[51] (594-95)

(But it is worth noting not everyone was busy visiting cultural sites; Strabo [6.2.8 (C. 274)] recounts the report of people he knew who had recently climbed Etna.)[52]

In our poet's opinion, not only do some others offer fictitious accounts of the world, and of natural phenomena such as Etna, but they also draw attention away from nature itself. His self-proclaimed task, as a poet, is to offer a true account of the cause of the volcano, and to encourage greater appreciation of nature. He urges his readers to "look upon the colossal work of the artist nature" (*artificis naturae ingens opus aspice*) (601). Yet even as the *Aetna* poet criticizes some poets—for retelling myths and legends and turning attention away from nature—he does not reject poetry as a medium of "scientific" discourse. On the contrary, his insistence on using poetry to present his natural-philosophical explanation underscores the recognized power of the genre for communicating scientific ideas and practice.

Our poet proclaims that his primary concern is to explain a natural phenomenon—in this case a particular volcano. But he is concerned not only with explaining Etna: throughout the poem he urges his readers to participate and engage in studying natural phenomena; he also offers numerous examples and hints regarding what is—in his view—good "scientific" practice, while also providing examples of what should be avoided.

At 222-251, the poet offers what is almost an "ode" to physical enquiry. He acknowledges that the task is demanding: "Infinite is the toil, yet fruitful too. Just rewards match the worker's task."[53] Humans, in his view, are required to seek understanding:

> Not cattle-like to gaze on the world's marvels merely with the eye, not to lie outstretched upon the ground feeding a weight of flesh, but to grasp the proof of things and search into doubtful causes, to hallow genius, to raise the head to the sky. (224-27)

He offers a list of various natural phenomena worthy of study, including the motions of the sun, moon, and stars, the nature of the seasons, and weather signs. But he emphasizes that the most important task for humans is "to know the earth and mark all the many wonders nature has yielded there" (252-53). This is, he stresses, "for us a task more akin than the stars of heaven" (254):

> For what kind of hope is it for mortal man, what madness could be greater—that he should wish to wander and explore in Jove's domain and yet pass by the mighty fabric before his feet and lose it in his negligence? (255-57)

Our poet bemoans the trivial matters with which most people are concerned, what he regards as a greedy preoccupation with overflowing barns and full wine-casks. Instead of being preoccupied with accumulating food and wealth, he urges that "everyone should imbue himself with noble accomplishments. They are the mind's harvest" (274-75). In his view (276-80), the goal of human beings should be

> to know what nature encloses in earth's hidden depth, to give no false report of her work, not to gaze speechless on the mystic growls and frenzied rages of the Aetnaean

mount, not to blench at the sudden din, not to believe that the wrath of the gods has passed underground to a new home, or that Tartarus is breaking its bounds.[54]

The aim of human endeavor should be to study nature: more specifically, "to learn what hinders winds, what nurtures them, whence their sudden calm . . . why their furies increase" (281-83). This is the ethical pitch of the *Aetna* poet: the pursuit of knowledge about the natural world, for its own sake, is a worthy goal.

The *Aetna* poet's call for the study of nature as a pursuit in and of itself contrasts with the self-proclaimed motives of the Epicureans, including Lucretius. The Epicureans argued that natural philosophy was useful only insofar as it promoted *ataraxia*. Epicurus himself warned that too much natural inquiry—such as that undertaken by some astronomers—could actually increase anxiety; astronomical phenomena were particularly problematic from Epicurus' point of view, simply because of the impossibility of viewing such phenomena up close.[55] The *Aetna* poet does not promise any other benefit from studying nature, other than the belief that it is a worthy occupation for humans. At the risk of being anachronistic, let me suggest that he seems to be calling for science for science's sake, and for pure (rather than applied) science.[56]

Nor is the poet content simply to advocate the study of the natural world: as he explains the true cause of Etna, he is careful to indicate to his readers the proper means to acquire understanding of the cause. The *Aetna* poet offers a natural-philosophical explanation of Etna, while advocating the pursuit of natural philosophy, and offering his own philosophy of natural philosophy, as it were, his own philosophy of science. And, of course, he does so through the medium of the poem.

Turning to the poet's explanation of Etna, we find that he was not engaged in what might be termed "volcanology." As Harry Hine has argued, the ancient authors who wrote on volcanic activity tended to focus on particular, local examples, rather than on any broader category. This is in contrast to, for example, the explanation of earthquakes, which were treated, as part of meteorology, as a more general occurrence, probably because they were also more frequently experienced in the relevant period and geographical areas.[57] (Etna has been more active in modern times than it was in antiquity.)

How does our poet explain Etna? The principal cause of Etna's eruptions, he writes, is wind, moving at high pressure in narrow subterranean channels; the volcanic fire receives nutrition from lava-stone (*lapis molaris*) (400-401).[58] Throughout the course of the discussion of the cause of Etna, our poet offers suggestions regarding how to develop proper scientific explanations. He recommends the use of observational evidence, and analogies drawn from everyday experience. Multiple explanations, which offer alternative possible causes, are also considered useful. The approaches he commends as part of sound scientific practice are very similar to those adopted by others who offered naturalistic explanations of meteorological and seismic phenomena. So, for example, Theophrastus and the Epicureans particularly favored the use of multiple explanations; similarly for these authors, wind was also cited as a cause of other phenomena.[59]

Our poet's explanation of Etna is not offered in isolation from his discussion of scientific method; he does not, for example, first outline his methodology, and then present his theory. Nor does he offer his explanation, and then justify it with an appeal to his philosophy of science. Rather, the poet weaves his discussion of methodology into his explanation, using parts of his explanation to serve as examples of proper procedure, while citing examples of what are in his view wrong-headed notions as warnings to his readers.

He begins by explaining that the earth is not completely solid: "Everywhere the ground has its long line of fissure, everywhere is cleft and, hollowed deeply with secret holes, hangs above narrow passages which it makes" (96-98). Drawing an analogy between earth and living organisms (98-101), he explains that

> as in a living creature veins run through the whole body
> with wandering course, along which passes every drop of
> blood to feed life for the selfsame organism, so the earth by
> its chasms draws in and distributes currents of air.

This allusion to the earth as living body is reminiscent of certain passages in Lucretius.[60] While it is not possible here to delve deeply into the question of our poet's philosophical orientation, there are several similarities between his scientific methodology and that of the followers of Epicurus' philosophy. So, for example, the *Aetna* poet suggests that there may be more than one explanation for the earth being the way that he describes it (102-15):

(1) "Either, I mean, when of old the body of the universe was divided into sea, earth and stars, the first portion was given to the sky, then followed that of the sea, and earth sank down lowest of the three, albeit fissured by winding hollows . . .";

(2) "Or maybe the cause of it is indeed ancient, though the formation is not coeval with its origin, but some air enters unchecked and works a road as it escapes";

(3) "or water has eaten away the ground with the mud . . .";

(4) "Or again hot vapours cribbed and confined have overcome solidity and fire has sought a path for itself."

Having offered these several possible explanations as to why the earth has fissures, the poet goes on to explain that, in his view, there is no objection to offering multiple explanations, for "no cause is here for mourning our ignorance, so long as the working of the true cause stands assured" (116-17).

Many of our poet's views on acceptable scientific practice resonate not only with the teachings of Epicurus, but also with the works of other authors, including Theophrastus and Seneca. The appeal to observation, like the use of multiple explanations, is often invoked by these authors, and by the *Aetna* poet, who asks: "Who does not believe that there are gulfs of emptiness in earth's recesses, when he sees springs so mighty emerge and so often plunge again in the depth of a chasm?" (117-19).

The *Aetna* poet repeatedly commends observation. Arguing that "confined winds have liberating vents which are concealed" (134-35), he emphasizes that "proofs of this through facts indisputable, proofs which hold the eye, the earth will give you in due order" (135-36). He invites his readers to make their own observations of hidden depths within the earth: "do you see how in forests there are lairs and caves of widely receding space which have dug far down their deep-sunk coverts?" (140-41).

Our poet recommends another method of scientific explanation used by many natural philosophers, including Aristotle: drawing analogies to everyday experience. He asks the reader to "let but your mind guide you to a grasp of cunning research: from things manifest gather faith in the unseen" (144-45). As an example, he suggests that

> as fire is always more unfettered and more furious in confined spaces, and as the rage of the winds is no less vehement there, so to this extent, underground and in earth's depths, must fire and wind cause greater changes, all the more loose their bonds, all the more drive off what blocks their course. (146-49)

But the poet's emphasis on the value of observation and analogy does not lead him to deny that some activities within Etna at times may be hidden from view. He cautions his readers against believing "the blockish rabble's falsehood" that Etna's power is at times diminished (367-69). Although we cannot always see what is happening, this does not argue against the cause. As with myth, he warns against believing popular, but false, views.

Our author's poem offers a powerful medium for communicating a range of scientific ideas. In many ways *Aetna* looks like a serious scientific text, describing and explaining a natural phenomenon, while offering instructions and advice regarding the best way to pursue the understanding of natural phenomena. The vivid language and imagery of the poem serve as the "honey" that Lucretius recommended, making the content of the poem more appealing. But in confronting the question of why the *Aetna* poet chose verse rather than prose, there are also indications that our poet wished his work to be regarded as belonging to a tradition, and one that was itself multi-layered.

At several points in the poem, the *Aetna* poet indicates his links to his predecessors, and to the tradition of didactic poetry. The didactic poem was a particular favorite of the Romans, with many examples dealing with scientific and technical subjects, including Virgil's *Georgics* (on agriculture), Ovid's *Fasti* (on calendars), and, as mentioned earlier, several Latin translations of Aratus' *Phaenomena* (on astronomy). From an ancient author's perspective, there was a good deal of evidence that this was a genre demonstrably attractive to a large readership; part of the appeal of the genre, as an authorial choice, may well have been knowledge that it attracted readers.

One of the key features of didactic poetry is that it self-consciously "draws attention to its status as poetry," through the poem's first-person speaker being identified with the poet himself, and references to his words as poetry or song.[61] Our author begins his work by announcing that it is a poem: "Aetna shall be my poetic theme"; "my poem shall tell" its causes.

Mount Etna was something of a *topos*, or literary commonplace, for poets in the ancient Mediterranean world.[62] Hine has noted that the earliest "proper" description of a volcanic eruption is found in one of Pindar's poems (*Pythian Ode* 1), dated to the late 470s BCE, describing Etna: "Etna, from whose inmost caves burst forth the purest founts of unapproachable fire."[63] Seneca, an author who wrote extensively on meteorology, including what we would today regard as seismology, mentions poems by Ovid, Virgil, and

Cornelius Severus touching on Etna.[64] From our author's point of view, if you were going to write about Etna, you would be in very good company indeed if you chose poetry as your medium.

However, in the *Aetna* poem, Etna is not only a *topos*, but is the subject of the work. Furthermore, the poet claims, in the opening lines, to be doing something new, in offering his explanation of the volcano. The poet seems to be seeking simultaneously to align himself with a tradition, while disassociating himself from some poetic traditions, and, importantly, embarking on what he claims is uncharted territory: the understanding of the causes of Etna. Even though other poets have written about Etna, he claims that his poetic project is a new one.

F. R. D. Goodyear, whose edition of *Aetna* established the standard, acknowledged that the subject of the poem was unusual within the body of known ancient texts; he noted that "apart from the 'Aetna' no known work nor any distinct portion (such as a book) of any known work by any Greek or Roman covered the subject of volcanism as a whole." Recognizing that while "what we happen to know about the literature of ancient Greece and Rome depends largely on the random chances of survival or of mention in surviving works," he argued that "it is very unlikely that a Latin poet opened a new field of scientific exploration in hexameter verse." Goodyear argued that the *Aetna* poet was to some extent responding to the ideas of others, "for he labours to refute views on volcanism divergent from his own and, as he believes, erroneous."[65] If the *Aetna* poet was writing in reaction to other texts, those texts are not now known to us; those other poems that mention Etna do so only briefly, without the sustained discussion and explanation offered by our poet.

Our poet uses poetry to criticize other poets' use of poetry; he is wary of some of the conventions associated with poetry, in particular the tendency for poets to offer mythologically based stories and explanations of natural phenomena. Nevertheless, he aligns himself with a particular tradition, that of Latin didactic poetry, and we see some echoes of other poets in his work. Lucretius also distanced himself from poetic and mythological traditions, while many Latin poets claimed to be doing something "new."[66] Siegfried Sudhaus and Friedrich Solmsen each pointed to borrowings in the opening lines of *Aetna* from both Lucretius and Virgil; Solmsen suggested that "the author makes it a point to introduce himself with bows of courtesy to the acknowledged masters of the didactic epic."[67] It seems clear that our poet wished his work to be identified in a tradition associated with the great names of Latin poetry.[68]

Even though the poet warned his readers about myths referring to gods, the *Aetna* ends with a legend about nature, man, and a god (Pluto). The poet concludes his work by recounting a moral tale about the virtuousness of nature in not harming two devoted brothers who sought to save their parents during an eruption. Nature, accordingly, shielded the family from harm, and the god Pluto (referred to in the poem by the name "Dis") allotted them a special place in the underworld (606-46):

> Once Aetna burst open its caverns and glowed white-hot
> ... a vast wave of fire gushed forth.... Two noble sons, Amphinomus and his brother, saw how their lame father and their mother had sunk down (alas!) in the weariness of age upon the threshold ... For *them* a mother and a father are the only wealth: this is the spoil they will snatch from the burning.... O sense of loving duty, greatest of all goods! Flames slant aside ... the greedy fire restrains itself. Unhurt [the family goes] free.... To [the brothers] the [poets] do homage: to them under an illustrious name has Pluto allotted a place apart. No mean destiny touches the sacred youths: their lot is a dwelling free from care, and the rightful rewards of the faithful.

Thus, the dutiful and loving sons save their aged parents from the fiery terror of Etna, and receive their reward from the god.

The *Aetna* poet was not the only one to sing the praises of these brothers. A number of Greek and Roman authors recounted their virtue, in poetry and prose;[69] statues were erected in their honor at Catina (the Latin name for what is now known as Catania, near Etna), and Sicilian and Roman coins were minted depicting them.

However, in spite of the brothers' high reputation, there is ambiguity here, in *Aetna*. Given that our poet has railed against legends and myth, it may come as a surprise to have this legend close the scientific account. Of course, our poet may be exercising his poetic license. Another explanation might be that, in a similar manner to how Seneca, a possible contemporary, closed his discussions of phenomena in the *Natural Questions* with moral messages, the *Aetna* poet also chose to end his work with a moral tale, even if that tale was a myth about a god. The *Aetna* poet is critical of myth and legend as a means to explain natural phenomena, yet ends his poem with a story about the benevolence of nature and a god rewarding human virtue.[70]

In this way, the *Aetna* poet delineates a boundary for the use of legend and the role of gods. It is not that the gods have no place in the world of the *Aetna* poet, but rather that their activities are circumscribed. At the beginning of his work, the *Aetna* poet invokes the favor of Apollo, as his muse, echoing Lucretius' opening prayer to Venus to inspire his poem. And the *Aetna* poet, like Lucretius, argues strenuously against the idea that gods can control or cause natural phenomena. These poets invoke the assistance of a god as muse, but reject any causative role for gods in the natural world. And, while legends may be useful to relate human-interest stories, only a rational and scientific approach serves to explain natural phenomena.[71]

In the following chapter, I will focus on the juxtaposition of "rational" and mythological explanations in a dialogue by Plutarch.

OF THE FACE APPEARING
in the roundle of the Moone.

Ell, thus much said *Sylla*, for it accorded well to my speech, and depended thereupon: but I would very willingly before all things else know, what need is there to make such a preamble for to come unto these opinions, which are so currant and rife in every mans mouth, as touching the face of the Moone. And why not (quoth I) considering the difficultie of these points which have driven us thither: for like as in long maladies, when we have tried ordinarie remedies, and usuall rules of diet, and found no helpe thereby, we give them over in the end, and betake our selves to lustrall sacrifices and expiations, to amulets or preservatives for to be hanged about our necks, and to interpretations of dreames: even so in such obscure questions, and difficult speculations, when the common and ordinarie opinions, when usuall and apparent reasons wil not serve nor satisfie us, necessary it is to assay those which are more extravagant, and not to reject and despise the same, but to enchant or charme our selves, as one would say, with the discourses of our auncients, and trie all meanes for to finde out the trueth: for at the very first encounter you see, how absurd he is & intollerable, who saith, that the forme or face appearing in the Moone, is an accident of our eie-sight, that by reason of weaknes giveth place to the brightnesse thereof, which accident we call the dazzeling of our eies, not considering withall, that this should befall rather against the Sunne, whose light is more resplendent, and beames more quicke and piercing, according as *Empedocles* himselfe in one place pleasantly noteth the difference, when he saith:

> *The Sunne that shines so quicke and bright,*
> *The Moone with dimme and stony light.*

for so he expresseth that milde, amiable, pleasant, and harmelesse visage of the Moone: and afterwards rendereth a reason, why those, who have obscure & feeble sights, perceive not in the Moone any different forme or shape, but unto them her circle shineth plaine, even, uniforme and full round about; whereas they who have more quicke and piercing eies, doe more exactly observe the proportion and lineaments, and discerne better the impression of a face, yea, and distinguish more perfectly and evidently the severall parts: for in mine opinion it would fall out cleane contrary, in case the weakenesse of the eie being overcome, caused this apparition, that where the patient eie is more feeble, there the said apparence and imagination should be more expresse and evident: furthermore, the inequalitie therein, doth fully every way confute this reason; for this face or countenance is not to be seene in a continuate and confused shadow: But *Agesianax* the Poet, right elegantly depainteth in some sort the same, in these words:

> *All round about environed*
> *With fire she is illumined:*
> *And in the middes there doth appeere,*
> *Like to some boy, a visage cleere:*
> *Whose eies to us doe seeme in view,*
> *Of colour grayish more than blew:*
> *The browes and forehead, tender seeme,*
> *The cheekes all reddish one would deeme.*

For intrueth darke and shaddowy things, compassed about with those that are shining & cleare are driven downeward, and the same doe rise againe receprocally, being by them repulsed, and in one word, are interlaced one within another, in such sort as they represent the forme of a face lively and natuturally depainted: and it seemeth that there was great probability in that which *Clearcus* said against your *Aristotle*. For this *Aristotle* of yours, though he familiarly conversed with that ancient *Aristotle*, perverted and overthrew many points of the Perepateticks doctrine. Then *Apollonides*, taking upon him to speake, demanded, what opinion this might be of *Aristotle* and upon what reason it was grounded. Surely (quoth I) it were more meet for any

man

The first page of Philemon Holland's translation of Plutarch's dialogue, in *The Philosophie, commonlie called, the morals* (London: 1603), is reproduced here by permission of the Syndics of Cambridge University Library (Ely.a.187). (This will be the subject of a further study, to be published separately.)

CHAPTER 3

Scientific and Mythic Explanation in Dialogue

Since antiquity, the recognized formats used to publish scientific work have not remained static. Some ancient formats, for example, the teaching text, continue to be used, while others have been discarded for scientific use, for example, poetry.[1] In this chapter, I will focus on dialogue, a genre particularly associated with certain authors, especially Plato. While Plato was not particularly concerned with natural philosophical and mathematical subjects in his dialogues, some later authors used the dialogue form to discuss such topics. In the early modern period, two authors interested in promoting Copernican cosmology turned to the dialogue. Galileo Galilei wrote two important "modern" scientific dialogues, on the *Two Chief World Systems* (1632) and *Two New Sciences* (1638). Johannes Kepler produced a Latin translation of a Greek work that much influenced him: Plutarch's (c. 45-after 120 CE) dialogue *On the Face on the Moon*, which was appended to his own work, the *Somnium* (Dream) (published posthumously, 1634).[2] Plutarch's dialogue has clear resonances with Plato's (c. 429-347 BCE) dialogue the *Timaeus*. Both of these ancient dialogues are concerned with "scientific" cosmology, and both incorporate myth as part of their cosmological explanation. It could be argued that, for Plato, the only medium that is suitable for offering a scientific account of nature is the myth recounted within the dialogue.[3]

This is particularly interesting, because it goes against some of our conceptions about what constitutes science. The idea that in the ancient Greek world there was a transition from myth to rational explanation—from *mythos* to *logos*—has been a powerful one. Some modern authors have referred to this alleged shift as a "miracle," while others have hailed a revolution, or even revolutions.[4] As was discussed in the first chapter, the suggestion that the emergence of philosophy depended on emphasizing *logos* (reason) and rejecting *mythos* (myth) is not a modern invention. A number of ancient Greek philosophers and other intellectuals sought to emphasize distinctions between what may be regarded as two different modes of explanation: the rational and the mythological. Some ancient thinkers—notably Plato—criticized poets and mythology.[5] And yet some ancient authors, including Plato himself as well as Plutarch, used both mythic and scientific explanations within the same work to address questions about the cosmos.

Here I will focus on Plutarch's dialogue *On the Face on the Moon*, while making some references to the *Timaeus*. I will ask whether the choice of dialogue as the genre is particularly relevant for Plutarch's cosmological discussion, in which rational (even mathematical) explanations are offered, as well as myth. In thinking about the genre of dialogue as scientific discourse, I will also ask: Why are rational accounts juxtaposed with myth? Are any claims being made about the relative value of rational and mythological explanations? What role does expertise serve in providing and assuring knowledge (scientific or mythological)?

Before turning to Plutarch's work, we should consider the genre of "dialogue" more generally, in order to have some idea of what this choice may have meant as an authorial option. The Greek noun *dialogos* has a range of meanings. While "conversation" or "dialogue" are the most prevalent, the meaning can range from "talk" or "chat" (Cicero *Letters to Atticus* 5.5.2) to "debating arguments" (Aristotle *Posterior Analytics* 78a12); furthermore, and especially because of the active character of philosophy, the associated verbs also display a range of meaning.[6] Etymologically, *dialogos* is related to *logos*, which itself has a wide range of meanings, including "computation or reckoning," "relation or correspondence," "explanation," "account," "debate," "narrative," "utterance," and "subject-matter." This richness of possible reference is reflected in the ancient examples of dialogues that survive.[7]

As a genre, the dialogue is often understood as a "special literary-philosophical form of writing" that had its origin in the philosophical activities of Socrates (469-399 BCE).[8] While others in Socrates' circle wrote dialogues, for example Xenophon (*c.* 430- *c.* 354 BCE), Plato's are normally regarded as influencing and shaping the entire genre. The dialogue form used by Plato generally conveys a sense of conversation, as well as argument and discussion of philosophical points; Socrates is often an important participant. While Plato chose the dialogue form for his publications, he never included himself explicitly as a speaker. Although some readers have argued that Socrates serves as Plato's mouthpiece, the extent to which this is the case is a matter of debate, and so it is not always clear which ideas contained in the dialogues were actually held by Plato himself. As Geoffrey Lloyd has noted, in the Platonic dialogues the speakers each have their own distinctive views and assumptions; this emphasizes the interactivity of the exchanges portrayed in the dialogue. Further, the dialogue represents a preferred mode of philosophizing, which is dialectical and presents different points of view, often through questions and answers.[9] The speakers in Plato's dialogues discuss a variety of topics, many of which have special importance for scientific enquiry, including questions relating to the nature of knowledge, the value of sense-perception, and the uses of language.[10]

In the *Poetics* (1447b11), Aristotle described written philosophical dialogues as "Socratic *logoi*," reflecting the association of the form of the dialogue with Socratic conversations. Conversation, and all that it might encompass, is understood as the basic format of the dialogue and, as we have probably all experienced, conversation can embrace a wide range of discourse and behavior; similarly, the genre may include anything from dinner-table talk amongst a group of participants to dialogues in which individuals offer very lengthy speeches—almost monologues.[11] While a typical Socratic dialogue emphasizes a question-and-answer format, not all of Plato's dialogues are set out this way; notably much of the *Timaeus*, as we have it, is a monologue, a lengthy discourse offered by Timaeus himself.[12]

Robert Lamberton has emphasized that for Plato "the Socratic dialogue was never an exercise in [simple] transcription"; the text does not represent a record of a conversation, nor does Plato claim to be reporting a discussion.[13] Rather, the choice of genre emphasizes philosophical enquiry, and in a rather dramatic way. The conversational format of the dialogue relies on the participation of a number of speakers, each with his own point of view and interests. Not only do the participants not always agree, the possibility of disagreement, debate, discussion, explanation, and other reactions is crucial to the interactions between interlocutors. This is an important feature of the genre: dialogue offers "the opportunity to juxtapose incompatible modes of explanation without sacrificing one to the other,"[14] allowing the author to present "multiple solutions to a given question and to keep them in balance, in tension with one another, without conceding that they are mutually exclusive."[15]

Together with Plato's contributions, those of Plutarch represent what survives of Greek philosophical dialogues. Aristotle's dialogues are no longer extant, except for some fragments. The Roman orator, statesman, and philosopher, Marcus Tullius Cicero (106-43 BCE) wrote a series of dialogues, presenting the principal positions of the leading Hellenistic schools on a number of philosophical and scientific topics; here he is an important, but not impartial, source of information regarding the ideas of many of the philosophers working after the death of Aristotle (in 322 BCE). Cicero sought to make Greek philosophy accessible to Roman readers, and he outlined the positions of various schools on a range of philosophical issues; the dialogue was a particularly convenient way to introduce different points of view to his readers. In the same way, the usefulness of the dialogue as a vehicle to explore divergent—even conflicting—ideas, without forcing the reader (or writer) to make a choice between them, must have motivated Plutarch in his choice of the genre for his treatment of cosmology. The dialogue offered Plutarch the possibility of approaching his topic from very different perspectives, without necessarily declaring his own views.

Plutarch (born before 50 and died after 120 CE) was born in Chaeronea in Boeotia in central Greece, studied philosophy in Athens, visited both Egypt and Italy, and lectured and taught in Rome. Known as a philosopher and biographer, he was also a priest at Delphi, and played an important role in the revival of the shrine. Many of his works are dialogues, written in the tradition of Aristotle rather than that of Plato, with long speeches and the appearance of the author himself. He is, however, regarded as a Platonist, and he was particularly interested in Plato's *Timaeus*. The participants in the dialogue *On the Face on the Moon* discuss the nature of the "face" that appears on the moon, as well as the role of the moon in the cosmos.[16] As in the *Timaeus*, the dialogue juxtaposes the ideas and arguments of natural philosophers and mathematicians with mythological accounts. The conversation and discussions reported in *On the Face on the Moon* include both the consideration of natural philosophical questions relating to the appearance of the moon, its size and material constitution, and also the recounting of what seem to be rather exotic stories (*mythoi*) about the moon—whether it is inhabited, its relationship to the soul, and its association with various gods and goddesses.

Plutarch's dialogue has a complicated structure. The beginning of the text appears to be mutilated, and it is not clear how much is missing. (There are two surviving manuscripts.)[17] The text can be understood as a dialogue within another dialogue. Both are recounted by the same narrator, Lamprias, who reports an earlier conversation to the participants of the larger, surrounding dialogue.

Some portions of the dialogue are presented as indirect discourse, others are quoted directly; these shifts add to the complexity of the narration.

The work, as we have it, begins with a reference to remarks made by Sulla, who is identified later in the text as a Carthaginian (942C). It ends with Lamprias, the narrator, recounting the words of Sulla, in which he retold a myth he heard from a stranger, who in turn had heard it from the chamberlains of Cronus (945D). (According to mythology, Cronus was the youngest son of Uranus (Heaven) and Gaia (Earth), and the father of Zeus.) The structure of the dialogue may be briefly outlined as follows:

(1) The narrator, Lamprias, quotes Sulla's opening reference to the story (*mythos*) he will later relate at the end of the dialogue;

(2) Lamprias reports an earlier "scientific" discussion about the moon, involving named contributors, and the ensuing conversation by participants in the larger dialogue;

(3) Sulla recounts the myth (*mythos*) that he heard from an unnamed stranger.[18]

The text begins in this way (920B):

> These were Sulla's words. "For it concerns my story [*mythos*] and that is its source; but I think that I should first like to learn whether there is any need to put back for a fresh start to those opinions concerning the face of the moon which are current and on the lips of everyone."[19]

Readers are given the clear impression that a discussion is already underway. But that the " fresh start" is made to report an earlier conversation only becomes completely clear well into the text (at 937C) when the narrator, Lamprias, states: "So we for our part, said I, have now reported as much of that conversation as has not slipped our mind."

The text indicates that many of the participants in the larger dialogue were not part of the earlier, reported discussion; the leader of the earlier discussion is not present in the dialogue *On the Face on the Moon*, but may have been meant to be Plutarch himself.[20] Harold Cherniss has contended that "that earlier discussion cannot, however, be identified with any that Plutarch may have had with his friends or with any lecture that he may have given; it is primarily a literary fiction, part of the structure of the dialogue for which it provides a specious motivation."[21] Nevertheless, the style of Lamprias' narration of both the earlier conversation and the larger dialogue—and his concern for what might have slipped his mind—give a sense of reportage. The reader is encouraged to think that the discussion may have actually taken place, amongst real people.

Sulla's opening words also reinforce the immediacy of the conversation. Although he only speaks very briefly at what is the beginning of our text, he emphasizes that the question of why the moon shows a "face" is widely debated: everyone is talking about it. These few words serve to bracket off the account of the earlier conversation concerning the face on the moon, which Lamprias then reports, offering a lengthy and detailed account. Once this recounting is finished, at the point when the larger dialogue resumes (937C), Lamprias announces that "it is high time to summon Sulla, or rather to demand his narrative as the agreed condition upon which he was admitted as a listener." This recalls Sulla's opening words, where he referred to "his story and its sources"; only then does the reader realize that Lamprias has recounted the earlier conversation on the promise that Sulla will relate a story about the moon.

In addition to the dialogue-within-dialogue format, the work may be read as being composed of two principal sections, each with its own style of explanation. The first part (920B-940F) is usually regarded as an exercise in natural philosophy, that is, as "scientific"; the second part reports a myth. The earlier conversation is a scientific debate about the nature of the moon and its place in the cosmos; within the larger, surrounding dialogue, scientific discussion continues before turning to the myth, which closes the discourse.

In the "scientific" section of the work, topics in astronomy, cosmology, geography, and catoptrics (that is, the study of mirrors) are discussed and debated. A number of different, sometimes conflicting, natural-philosophical and mathematical explanations of the appearance of the moon's "face," the source of the moon's light, the material constitution of the moon, and whether it belongs to the terrestrial or the celestial region of the cosmos are presented. Rival views associated with the several different philosophical schools (the Academy, the Lyceum, and the Stoa) are aired. The principal participants are Lamprias and Lucius, well-educated men but not specialists. Nevertheless, the standard of conversation is sophisticated, and specialist, technical works are quoted (including Aristarchus' *On the Sizes and Distances*).

So, for example, Lamprias reports on the earlier conversation, in which he gave an account of the Stoic philosopher Clearchus' explanation of the face on the moon:

> what is called the face consists of mirrored likenesses, that is images of the great ocean reflected in the moon, for the visual ray when reflected naturally reaches from many points objects which are not directly visible and the full moon is itself in uniformity and lustre the finest and clearest of all mirrors. Just as you think, then, that the reflection of the visual ray to the sun accounts for the appearance of the <rainbow> in a cloud where the moisture has become somewhat smooth and <condensed>, so Clearchus thought that the outer ocean is seen in the moon, not in the place where it is but in the place whence the visual ray has been deflected to the ocean and the reflection of the ocean to us.[22]

This is one of many explanations that are debated by members of the group; it presumes a degree of familiarity with contemporary science, including optics and meteorology. The discussants do not shy away from mathematical arguments, as when later in the conversation Lucius, speaking particularly to the mathematician in the group, reportedly said:

> In your presence, my dear Menelaus, I am ashamed to confute a mathematical proposition, the foundation, as it were, on which rests the subject of catoptrics. Yet it must be said that the proposition, "all reflection occurs at equal angles," is neither self-evident nor an admitted fact. It is refuted in the case of convex mirrors when the point of incidence of the visual ray produces images that are magnified in one respect; and it is refuted by folding mirrors, either plane of which, when they have been inclined to each other, and have formed an inner angle, exhibits a double image, so that four likenesses of a single object are produced, two reversed on the outer surfaces and two dim ones not reversed in the depth of the mirrors.[23]

The discussion is thoughtful and reflects the high level of education of the participants; various theories relating to the moon are recounted. Numerous well-known authorities and their texts, including the philosopher Posidonius (*c.* 135-*c.* 51 BCE) and the astronomer Aristarchus, who observed the summer solstice in 280 BCE, are cited and in some instances quoted.[24] The scientific section is challenging in its detail, requiring careful attention in order to be understood, and presuming, in places, a familiarity with other scientific texts. (While there are no direct references to diagrams, one can well imagine that the participants in the dialogue, and readers, would have found them helpful.)[25]

Some of the scientific section (to 937C) is the report of the earlier conversation. After recounting this, Lamprias indicates that the time has come to hear Sulla's promised story. He mentions that up to that point the participants in the dialogue had been out walking; he suggests, "if it is agreeable, let us stop our promenade and sit down upon the benches, that we may provide him with a settled audience" (937C-D). The participants then are seated, with the plan of hearing Sulla speak. With this change in setting and posture—a further indication of the presumed contrast between science and myth—Lamprias asks Sulla to begin, but before he has the chance, Theon, who is treated as the literary expert of the group,[26] puts in a special request: to hear about the beings that are said to dwell on the moon—not whether any really do inhabit it, but whether habitation there is possible.[27] He then offers his own views, sprinkled with references to Plato, myth, astronomy, and meteorology. His choice of topic is praised by Lamprias (at 938C), and the detailed consideration of the habitability of the moon continues, until Lamprias reports Sulla interrupting his disquisition about the moon (at 940F) to tell the story.

This interjection may be seen as marking out the final part of the work, which is largely devoted to an eschatological myth dealing with the question of what happens at death.[28] The myth describes the role of the moon in the cosmos: the moon is associated with the soul, and is the place to which souls go when they have left their bodies after death (945A) or have not yet been born into their earthly bodies (943A).[29] Sulla's sharing of the myth is presented as a somewhat delayed climax within the structure of the work.

Sulla refers to his account specifically as a *mythos*, and he makes it clear that he is simply reciting. He is not the author of the story he tells, but heard it from an unnamed stranger who had himself learned of it, as previously noted, from "the chamberlains and servitors of Cronus" (cf. 945D; 942C-945D). Sulla relates the stranger's account of how he had traveled to Carthage, where he and Sulla met.[30] Then, in the geographical introduction (941A-942C) that precedes the mythic account of the role of the moon, the stranger explained to Sulla how he had come to the isle of Cronus from a land far over the sea, west of Britain. (Kepler believed that this was America.)[31] And the stranger was very busy during his travels; Lamprias reports that Sulla related that "all his [the stranger's] experiences and all the men whom he visited, encountering sacred writings and being initiated in all rites—to recount all this as he reported it to us, relating it thoroughly and in detail, is not a task for a single day." Sulla asked only that we "listen to so much as is pertinent to the present discussion" (942B-C).

Lamprias' report of Sulla's recounting of the stranger's myth is lengthy and detailed (as was his report about the earlier conversation about the moon's face), and makes reference to philosophers (including Plato and Xenocrates, head of the Academy from 339-314 BCE, at 943F) and to the work of geometers (944A), as well as to the Homeric poems and numerous figures of myths. Various aspects of the moon, including its material composition (943F-944A), are explained, especially its special status and role in our cosmos. Sulla reported the stranger's words (942C): "among the visible gods he said that one should especially honour the moon, and so he kept exhorting me to do, inasmuch as she is sovereign over life." The moon is responsible for the human soul. The stranger explained that "most people rightly hold man to be composite but wrongly hold him to be composed of only two parts" (943A); rather, in the composition of the parts of man, "earth furnishes the body; the moon the soul, and the sun furnishes the mind for the purpose of his generation even as it furnishes light to the moon herself."

The moon's most important role (for humans) is as the resting place of the soul, when it is not within a body. So, after death "the substance of the soul is left upon the moon and retains certain vestiges and dreams of life as it were" (944E-F); "the moon is the element [of souls], for they are resolved into it as the bodies of the dead are resolved into earth" (945A). The dialogue ends, as it begins in the extant version, with the words of Sulla: "This I heard the stranger relate and he had the account, as he said himself, from the chamberlains and servitors of Cronus. You and your companions, Lamprias, may make what you will of the tale."

As Cherniss has pointed out, "hearing it from Lamprias now, the reader has [the mythological] part at fourth hand and the geographical introduction of the stranger at third hand";[32] there is a question of whether this distance from the original source is significant for Plutarch in his construction of the account. How are Lamprias' listeners and Plutarch's readers meant to form an opinion of what they have heard and read? Plutarch, through Lamprias, at various points provides information about the sources of scientific theories and myth.

The sources of the explanations offered

Some of the participants in the larger dialogue, including Lamprias himself, took part in the earlier conversation; others did not. At certain points there is ambiguity about the overlap of participation; whether this ambiguity is accidental is not clear.[33] Lamberton has suggested that "the slippage between ill-defined past and featureless present is constant and serves to throw into relief the arguments themselves (or reports of arguments)."[34]

Even though the reader is not always certain who participated in the earlier debate, the narrator provides information about the points of view and background of the various individuals named in the dialogue. This information about philosophical affiliation and expertise enables the auditors of Lamprias' narration of the earlier conversation (taking part in the larger dialogue), as well as the readers of the work as a whole, to form their own opinions about the discussions. The sources of the cosmological ideas—both the scientific theories and the myth—are identified, presumably with the view to helping to assess those ideas. Portions of the earlier conversation are reported as indirect discourse (for example, at 933F); most is presented directly, as an ongoing exchange. Some individuals are identified by their area of expertise, others by their allegiance to particular philosophical schools—for example, that of the Stoics. Lamprias, as the narrator of the entire work, mostly speaks in the first person; he also quotes those who took part in the earlier discussion. He is knowledgeable about various philosophical positions, criticizing Stoic doctrine and supporting positions identified with the Academy (922F); he may be taken to represent the educated layman.[35]

The background of some of the individuals named in the dialogue is indicated by the part they play in discussion, or by overt references. The other principal speaker in the earlier conversation, Lucius, sets out the theory of the moon (922F-923E) associated with the Academy. Apollonides is identified as someone who is knowledgeable about geometry (920F), whose area of expertise coincides with that of Hipparchus (the second-century BCE astronomer; cf. 921D and 925A). Menelaus, referred to as a mathematician, is addressed by Lucius at one point (930A), but he does not reply then nor does he speak elsewhere in the surviving parts of the dialogue. The speaker referred to as Aristotle (928E) represents the Peripatetic school, but is not taken to be the philosopher himself; Pharnaces is a Stoic.[36] Some speakers are identified more subtly through the interaction of the conversation, as when Sulla refers to Carthage as his country (942C).[37] Sulla's main role in the dialogue is to recount the myth, though it is not certain whether he should be regarded as a mythographer; he is not credited by Plutarch here with collecting other myths, and his status as a purveyor of myth is ambiguous. Some of the participants in the dialogue are learned experts in relevant subjects, such as astronomy; others, like Sulla, are presented as having acquired their information almost by chance. At certain points in the dialogue, specialist, expert knowledge is highly valued; at others, the reader is left wondering how to judge the sources of information and explanations on offer.

What is the relationship between the scientific and the mythic in this dialogue? Does the juxtaposition here of rational and mythological explanations suggest anything regarding the relative status of these two seemingly contrasting explanatory approaches? Are we meant to have a view regarding the status of explanations offered by experts, educated laymen, and strangers?

Here it is useful to turn to Plato. Much of the content of Plato's dialogues can be described as attempts to solve problems, in a rational and analytical manner. However, in a number of Platonic dialogues myth also plays an important part. In some instances the myth seems to take over, offering, as it were, another way of describing the world, an alternative to analytic discourse.[38] But the status of the myth is not always made clear, and is sometimes ambiguous. Plutarch was certainly familiar with Plato's dialogues; the inclusion of the myth in the dialogue *On the Face on the Moon* very likely was inspired by Plato's own incorporation of myth, particularly in his cosmological dialogue, the *Timaeus*; at certain points in Plutarch's dialogue, the *Timaeus* is glimpsed in the background.

The mythic and the scientific in the *Timaeus*

While it is usual to refer to the *Timaeus* as a dialogue, the work known to us is, largely, an extended speech given by Timaeus. The *Timaeus* is thought to have been written towards the end of Plato's life, roughly in the 350s BCE (he died in 347). It is part of the larger *Timaeus-Critias*, in which the members of a group—which includes Socrates—each speak about a specific subject. At the beginning of the *Timaeus* we learn that Socrates presented his own account of an ideal city, and then asked Critias to describe such a city in action; in the distant past (9000 years earlier), Athens itself was such a city as that described by Socrates. Critias is meant to recount the story of Athens' great war with Atlantis, but before he has his turn another participant, Timaeus, agrees to give an account of the origin of the cosmos; the remainder of the work is taken up by Timaeus' speech. The evidence of the surviving work indicates that Plato had intended for other speakers to have their say as well, on their own topics. It seems likely that Plato did not live to complete the entire work, but published the discourse known as the *Timaeus* as it stands.[39]

Critias (*Timaeus* 27a) explains to the others present that Timaeus "knows more of astronomy than the rest of us and has made knowledge of the nature of the universe his chief object; he will begin with the birth of the world and end with the nature of man."[40] Timaeus is introduced as a specialist with expert knowledge. As he begins his account of the cosmos and its origin (29c-d), he famously addresses Socrates and warns that if

> in many respects concerning many things—the gods and the generation of the universe—we prove unable to render an account at all points entirely consistent with itself and exact, you must not be surprised. If we can furnish accounts no less likely than any other, we must be content, remembering that I who speak and you my judges are only human, and consequently it is fitting that we should, in these matters, accept the likely story and look for nothing further.[41]

Timaeus presents an account of the origin of the world; it is an account of creation presented as a "likely story," with many mythic elements.[42] He rejects the possibility of arriving at the truth. To attempt to search for a completely true account of the world would be a futile exercise, for human beings can hope only to offer an explanation that is plausible.

Timaeus' speech can be read as a creation myth, in which the Demiurgos, the craftsman, fashions the world, its contents and inhabitants. As David Sedley has explained: "The creation story is obviously a myth. And myths are frequently used by Plato to convey important truth in a non-literal fashion." In Plato's dialogue *Protagoras*, the main interlocutor "undertakes to give his own demonstration that virtue is teachable, indicating to his audience (*Protagoras* 320c2-4) that the very same lesson could be expounded either with a simple discourse (*logos*) or with a myth (*mythos*)." Choosing the latter, Protagoras "illuminates the nature of our innate moral instincts by an entertaining creation story" that can be variously interpreted. As Sedley has noted: "To interrogate Protagoras' myth by asking exactly what it is saying about the origins of human society" misses the point, for "mythical discourse has ... indeterminacy as an inherent feature."[43] The use of myth as an explanatory tool makes use of this element of indeterminacy.

The creation account in the *Timaeus* includes numerous shifts in register throughout the discourse, between myth, fable, prayer, scientific analysis, philosophical argument, and explanation.[44] A. E. Taylor, in his 1928 commentary, argued that the cosmology of the *Timaeus* "properly speaking ... is not 'science' but 'myth,' not in the sense that it is baseless fiction, but in the sense that it is the nearest approximation which can 'provisionally' be made to exact truth."[45] The choice of myth as the vehicle for the explanation enables Timaeus to emphasize the provisional character of his account. As Sedley has explained, "Timaeus invokes a strong version of the Platonic 'two world' metaphysics, which separates a realm of intelligible being from one of perceptible becoming. The physical world belongs to the latter realm, and must be explained with the kind of discourse appropriate to that realm, one aiming for 'likelihood' rather than ... cast-iron certainty."[46]

The account offered by Timaeus also incorporates detailed cosmological ideas and arguments, as well as mathematical explanations. While it is not possible to consider the text in any detail here, Plato's creation myth includes serious scientific argument. The discourse is simultaneously "scientific" and "mythic"; it would be difficult and misleading to suggest that there is a clear-cut division in the *Timaeus* between these two possible modes of explanation. Timaeus' description of the way in which the Craftsman constructed the circles in the World-Soul is indicative (36b-c):

> This whole fabric, then, [of the World-Soul] he [the Craftsman] split lengthwise into two halves; and making the two cross one another at their centres in the form of the letter X, he bent each round into a circle and joined it up, making each meet itself and the other at a point opposite to where they had been brought into contact.
>
> He then comprehended them in the motion that is carried round uniformly in the same place, and made the one the outer, the other the inner circle. The outer movement he named the movement of the Same; the inner, the movement of the Different. The movement of the Same he caused to revolve to the right by way of the side; the movement of the Different to the left by way of the diagonal.[47]

Commentators since antiquity have sought to explain this passage, understood as astronomical and cosmological, in which Timaeus describes the fashioning of the celestial sphere by the Craftsman. The circle of the Same is considered as the motion of the World-Soul, and of the sphere of the fixed stars; the circle of the Different is associated with the motions of the planets.

Many readers would have been mystified by Timaeus' terminology; Plutarch (*De audiendo* [*On Listening to Lectures*] 43a-b) refers to young men showing off, who display their mathematical knowledge speaking about problems relying on the meaning of "by way of the diagonal." F. M. Cornford has suggested that Timaeus was describing an astronomical model used for teaching, a sort of armillary sphere.[48] That there is mathematical content, and "scientific" intent, here is not disputed; what is interesting is the way that Timaeus' science is conveyed: his astrophysics is presented as part of a myth.

The situation is rather different in Plutarch's *On the Face on the Moon*. Here the science and the myth both play important but rather separate roles in the discourse. The perception that the two approaches—of science and myth—are completely separate has been reflected in the way the work has been studied by modern scholars. Most have paid more attention to the mythological elements of the dialogue, a situation bemoaned by Cherniss, who expressed the hope that the serious "scientific" content would warrant closer attention.[49] But this tendency to treat Plutarch's myth separately from his science belies the fundamental character of the work: it is a dialogue, in which both myth and science play significant roles, offering ways to understand the cosmos, and the character and nature of the moon.

Because the beginning of the dialogue is no longer extant, it is impossible to know what is missing. In the surviving portion, as explained earlier in this chapter, the discussion of the various scientific and mathematical explanations of the nature of the moon is framed from the outset by a reference by Sulla to the myth he recounts in the final section. In the opening lines he is quoted by Lamprias as having said (920B): "For it concerns my story [*mythos*] and that is its source; but I think that I should first like to learn whether there is any need to put back for a fresh start to those opinions concerning the face of the moon which are current and on the lips of everyone." We have seen these lines before, but now let us consider Lamprias' own response (920B-C):

> What else would you expect us to have done, . . . since it was the difficulty in these opinions that drove us from our course upon those others? As people with chronic diseases when they have despaired of ordinary remedies and customary regimens turn to expiations and amulets and dreams, just so in obscure and perplexing speculations, when the ordinary and reputable and customary accounts are not persuasive, it is necessary to try those that are more out of the way and not scorn them but literally to chant over ourselves the charms of the ancients and use every means to bring the truth to test.[50]

Suggesting that a frustration with the usual accounts of the face of the moon spurred him on, Lamprias launches into the discussion of various scientific explanations of the figure seen in the moon, including a weakness of our vision (920C).

The dialogue *On the Face on the Moon* is presented as the continuation of an earlier conversation; this suggests a direct parallel to the *Timaeus*, also explicitly set out in that way. In the case of the *Timaeus*, we are given what is to be assumed to be the full story about the origin and organisation of the cosmos, in other words, both a cosmogonical and a cosmological account. Even though the entire *Timaeus-Critias* was not finished, Timaeus' account seems to be complete, even if it is presented only as "likely." In *On the Face on the Moon*, the scientific portion is presented as a résumé of an earlier discussion; there are strong hints that there is more that could be told: it is not intended as a complete explanation, or the last word on the subject.

Not only is the scientific discussion juxtaposed to the myth; by one reading, the scientific discussion is *completed* by the myth that closes the dialogue, for it is the notion that the moon is inhabited by human souls that establishes the purpose of the moon in the cosmos. The myth allows the rounding off of the teleological message of the dialogue: the moon is earthy in its constitution; it is in principle habitable and therefore serves a purpose as the home of souls without bodies.

To conclude: Plutarch's dialogue *On the Face on the Moon* and Plato's *Timaeus* both incorporate two different styles of explanation: one usually considered to be scientific, the other based on mythology, although structurally, within each dialogue, this juxtaposition of the scientific and the mythic is accomplished differently.[51] In the *Timaeus*, the creation myth itself contains detailed natural philosophical and mathematical elaboration; in *On the Face on the Moon*, the scientific discussion is bracketed off and followed by an account of a myth that justifies the existence, teleologically, of the astronomical body whose nature and phenomena are the subject of the scientific discussion. Lamberton has argued that Plutarch's dialogues simultaneously proclaim "the seductive necessity of inquiry and our remoteness from both certainty and truth."[52] In the case of the dialogue *On the Face on the Moon* there is the implication, as in the *Timaeus*, that complete understanding and the whole truth cannot be gained through scientific means alone. For both Plato and Plutarch, myth also plays an important part in explaining the cosmos.

The Choice of the Dialogue

Let us return to the question of why Plutarch chose to present his cosmological ideas using the genre of dialogue. Would not it have been easier, and less complicated, to offer a simple prose treatise? The definition of genre involves identifying form and function as characteristic of particular types of discourse. The dialogue *form* allows the setting of reason and myth side by side; one *function* may be to present the reader with different epistemological models, different ways of knowing and understanding. Science and myth are in dialogue in Plutarch's *On the Face on the Moon*.

In Plutarch's dialogue, scientific enquiry and mythological explanation are not set up as rivals; rather, they are presented as two complementary aspects of a full consideration of nature. While expert knowledge is recognized and plays an important role in the discussion, the significance of a myth transmitted over distances of time, space, and culture is not derided. Furthermore, as is emphasized, we have not heard the full range of possible explanations until we have heard the recounting of the ancient myth.

The special status of dialogue, and its pertinence for understanding this juxtaposition of *logos* and *mythos*, has been emphasized by Hans Georg Gadamer, the leading exponent of philosophical hermeneutics, concerned with the methodology of interpretation. Gadamer's work engages closely with that of Plato and Aristotle, in particular. He suggested that "to reach an understanding in a dialogue is not merely a matter of putting oneself forward and successfully asserting one's own point of view, but being transformed into a communion in which we do not remain what we were."[53] Gadamer asserted the primacy of conversation over other forms of discourse;[54] in his view, this primacy of conversation and of dialectic is exemplified in the Platonic-Socratic dialogues. He emphasized that the art of dialectic is not the art of argumentation, but the art of questioning and seeking the truth.[55] In Gadamer's reading of Plato, truth is not only to be found in argument; myth is also a fundamental source of truth.

Gadamer celebrated what he regards as Plato's mythical refutation of dialectical sophism in, for example, the *Meno*, where sophistical objection is not "overcome there through superior argument, but by appealing to the myth of the pre-existence of the soul." Gadamer explained "this is a very ironic appeal, since the myth of pre-existence and anamnesis [recollection], which is supposed to solve the mystery of questioning and seeking, does not present a religious certainty but depends on the certainty of the knowledge-seeking soul, which prevails against the emptiness of formal arguments." He contended that "it is characteristic of the weakness that Plato recognizes in the *logos* that he bases his critique of the Sophists' argument not on logic but myth." But Gadamer acknowledged that "however convincing it seems, Plato's mythical refutation . . . does not satisfy the modern mind."[56]

This acknowledgment that "the modern mind" is not satisfied or content with the employment of *both* mythological accounts and rational arguments—much less the possible "triumph" of myth in a rational setting—highlights what is distinctively "ancient" about Plutarch's *On the Face on the Moon*. In Galileo's dialogues, there is clearly a "correct" position (that is Galileo's own), as well as a wrong one, which is not only criticized, but ridiculed (this is the point of view represented by the backward Aristotelian, Simplicius). Argument and the presentation of one's "correct" ideas are the purpose of Galileo's dialogue; the dramatic setting, presumably, is meant to engage readers and also possibly to protect those asserting the "truth."[57]

Plutarch's account is presented as a conversation, to which each participant is invited to contribute in a particular way. There is absolutely no expectation or prescription that there should be, or will be, agreement; rather, there will be questioning, and further thinking. That Plutarch recognized the power of myth is indicated by a statement, preserved in a fragment from a now-lost work *On the Festival of Images at Plataea*, that "ancient natural science, among both Greeks and foreign peoples, took the form of an account of nature hidden in mythology, veiled for the most part in riddles and hints, or of a theology such as is found in mystery-ceremonies: in it what is spoken is less clear to the masses than what is unsaid, and what is unsaid gives cause for more speculation than what is said."[58] In *On the Face on the Moon*, Plutarch intends that we are to continue to consider not only the different accounts of the nature of the moon—mythological, philosophical, and mathematical—but the ways in which we gain knowledge itself.

Writing about the *Timaeus*, Thomas Johansen has distinguished four senses in which the work is a dialogue: the dialogue that occurs between characters, the framing dialogue in which Timaeus' speech occurs, the dialogue between author and characters, and the dialogue with readers who encounter the work.[59] These four senses of dialogue apply equally well to *On the Face on the Moon*, which also utilizes a frame in its surviving literary structure.

Both Plato's *Timaeus* and Plutarch's *On the Face on the Moon* are strong evidence that dialogue has a special status as a genre of Greek scientific writing, in that it can provide a vehicle for the deliberate bringing together of scientific and mythic explanations. That is the purpose of the juxtaposition here, and the motivation for choosing the dialogue form: to engage us as readers in actively considering the ways in which we claim to know, to understand, and explain.

EPILOGUE*

The preceding three chapters began life as lectures. During and following the delivery of the lectures at Oregon State University, they benefited greatly from the comments, questions, and interjections of the original auditors; the texts of the lectures (and subsequent revisions) were then circulated to a number of readers, who offered their own very helpful and, sometimes, provocative suggestions. In some cases, I discussed the preceding chapters in conversations; in other instances, the discussion utilized the modern epistolary form of email. In all cases, I was reminded of the still somewhat fluid and permeable boundaries between oral and written discourse—so significant for the ancient texts I have considered here—as well as the ever-important interactions between "author" and "reader."[1]

I have deliberately chosen not to try to engage in the preceding chapters with issues relating to readership in antiquity.[2] Rather, I have been primarily interested in what might be termed "authorial choices," arguing that ancient authors writing about scientific and mathematical topics did have a range of genres from which to choose for the expression of their ideas. However, the choice of genre would have provoked responses from potential readers of those texts. Some may have decided against engaging with a text in a particular genre, while others may have been attracted to a particular format. That certain genres would have special appeal to readers was certainly assumed by some authors; Lucretius' suggestion that the "honeyed cup" of poetry would attract more potential readers than prose has already been mentioned.[3]

---

* "Epilogue"—from the Greek verb *epilegein*: to say more

Now I wish to consider another aspect of the genres and texts we have considered.[4] Though they have not been discussed here explicitly, readers and reading practices are important for the consideration of genres of scientific texts. While in many cases we do not know much about the readership of specific works, in others we do have some idea of who might have—or actually did—read certain works. For example, some works are dedicated to a particular addressee; while that person may not have actually read the work himself, he can—to some extent—be understood as the "projected" reader. In some cases, for example that of commentaries, we have clear evidence of how individual readers interacted with a given text. Some "reading" practices actually impacted on the texts we now have; for example, the editing and publication in the ancient period of the lectures of philosophers and teachers and the production of commentaries on other texts required a very active form of reading, as well as writing.[5] Recognition of the involvement of readers in the production of texts helps to focus our attention on the community of authors, auditors, editors, translators (as in the case of Greek texts translated into Latin),[6] and readers involved in the various aspects associated with the production, reception, and use of those texts.

Here, my aim has been to draw attention to the choices made by ancient authors seeking to share explanations of natural phenomena. I have chosen to focus on two genres—the poem and the dialogue—but I recognize that there may be difficulties in characterizing the texts highlighted here—the *Aetna* poem and Plutarch's dialogue *On the Face on the Moon*—as works of either science or literature. Indeed, I think it important to recognise that these texts—from our modern perspective—straddle what some readers might expect to be a boundary between the (natural) philosophical and literary text. (As was noted earlier, particularly because of the influence of Plato, the dialogue is sometimes referred to self-consciously as a literary-philosophical genre.)

Furthermore, it is difficult to know how to characterize the authors of these texts. Today, we normally refer to the author of *Aetna* as "the *Aetna* poet," recognizing his art, but not his natural philosophy. Plutarch—while acknowledged as an author with a strong philosophical background and orientation—is also generally appreciated more as an essayist than as a "philosopher."

But whatever the label we might attempt to attach to these authors, it is clear that they intended their work to be read. In choosing a particular genre, each author may have been aiming his work towards a particular type of reader—if indeed "types" of readers is a reasonable way to characterize the intellectual and literary marketplace for such works. Certainly, these works were intended for audiences already familiar with such genres and texts; there is the clear expectation that their audiences will be familiar with the genres of didactic poetry and philosophical dialogue.

By selecting a genre of communication, each author placed himself within a larger community—a community that included, on the one hand, other authors working in the same genre, and on the other, their readers or auditors. Some scholars would regard the genre as constituting a tacit contract between the author and the reader.[7] The sense of membership within a larger community is reinforced in a number of ways in these texts, through which the authors and their audiences form intellectual communities, sharing, considering, debating—in some cases, correcting or refining—ideas.

Not all reading would have been done as a solitary activity by individuals. David Konstan has pointed to the "abundant evidence that the ancient Greeks and Romans read—or were read to—in groups, under conditions that promoted commentary and conversation." He has cited Plutarch's recommendation in *How a Youth Should Listen to Poems*, where young people are encouraged to "respond to statements by classical poets which they regard as inappropriate, as though they were speaking directly to the author."

As an example, "when a character in a play by Sophocles affirms that 'profit is pleasant, even if it comes from falsehoods' (fragment 749), Plutarch blurts out (21A): 'but in fact we heard you say that "false statements never bear fruit" [fragment 750]'." But this "latter statement is not necessarily Sophocles' true opinion, according to Plutarch"; rather, Konstan explained, he is aiming "to encourage young readers to adopt a critical attitude toward all literature."[8]

William A. Johnson has argued that readings performed out loud in a group formed an important part of educated, elite male culture in the Greco-Roman world: "'Reading' in this society is tightly bound up in the conversation of the community. Group reading and serious conversation devolving from reading are twin axes around which much of the elite man's community turns." Reading and discussing texts together in a group served special social, as well as intellectual, functions, allowing the participants to form a community.[9]

For some, there may have been an additional attraction of participating not only in a community, but also in a tradition, which may itself be reinforced by a particular genre; the commentary, for example, reinforces the sense of community through participation in a tradition of active reading. One of the indications of an assumption of a community of authors and readers is the occurrence of what literary scholars refer to as "intertextuality": in the texts considered in the previous chapters, there tend to be frequent references to and echoes of other (especially earlier) authors and texts. In some cases these echoes of earlier works would have been easily recognized; in other cases the resonances would have been more subtle, and more demanding of the reader.

This intertextuality is demonstrated—to some extent—in the choice of genre, and the use of the conventions appropriate to that genre. Greek and Roman authors writing on scientific topics often refer to, discuss, criticize, and make use of the ideas, observations, and methods of others, particularly predecessors. Plutarch, like many philosophical authors writing in prose, often provides the name of the thinker associated with certain ideas, sometimes regarding these named individuals as authorities and experts, while treating others as proponents of a view to be argued against. But it appears to be a poetic convention that authorities are rarely cited by name, even when there are clear allusions to particular authors and texts.

A certain amount of arguing against one's predecessors takes place in both *Aetna* and in Plutarch's *On the Face on the Moon* and, as in some other texts, there is a degree of self-consciousness about this. As we have seen, in some cases the proper methods for observing and evaluating others' observations were considered and discussed in a very deliberate way. One point I should perhaps stress is that often (usually) the work that is referred to is the work of predecessors, some of whom have been dead for centuries. In fact, authors use the term "the ancients" to refer to some of their predecessors, highlighting the antiquity of their views (and in some cases their authority). Although dead, the ancient authors are regarded as having a living voice. Some authors were at pains to provide historical material about their predecessors, and even to preserve rare texts, as in the case of portions of Parmenides' poem preserved by Simplicius.[10] Who were these authors writing for? For readers, some of whom might live long after themselves, just as some authors lived long after their own predecessors with whose texts they actively engaged.

If we assume that these authors and their readers identified themselves to some extent as members of a community—at least of a community whose interests intersected around interactions with particular texts—certain questions should be addressed, even while recognizing that it is difficult and unwise to generalize across a long time span and vast geographical spaces. What is an intellectual (or even a "scientific") community? Who is a member? What function does the community serve? Do all of the members have to have the same intellectual orientation? For example, would they all be identified as scientists/natural philosophers, or might some of them be more interested, for example, in literature? Questions that particularly intrigue me are: Do all of the members have to be alive at the same time, in order to participate? Or can long-dead members be perceived as "active" members of the group?

In thinking about the constitution of an intellectual community, several other points should be raised. A community is distinguished from other groups within society by a sense of common purpose.[11] By positioning themselves as members of a "community," ancient authors emphasized that they were involved in a group effort or project to explain natural phenomena. Some of the ancient authors on scientific and mathematical subjects presented themselves as members of a particular community, for example, a philosophical school. The philosophical schools provided a strong sense of membership within a "community," sometimes very practically, with people interacting face-to-face. But in some cases membership in a philosophical community was at far remove, and was experienced through an identification with the proponents of particular philosophies, and the texts that conveyed those philosophies. In many cases, as noted previously, membership in a philosophical school conveyed a sense of membership not only in a community, but also of operating within a particular tradition.

The membership of an intellectual "community" could include competitors; members of the community need not always be in agreement. Geoffrey Lloyd has emphasized the agonistic side of ancient philosophy and science. That it is impossible to have competition if there is no one against whom to compete provides another argument for understanding a "community" as including even those who may disagree with a particular point of view or approach, but who nevertheless engage in discourse with a view to providing answers to a shared or related set of questions or issues.

Today we place a high premium on innovation. As a society we sometimes seem to think that innovation and "progress" should characterize scientific work; indeed, we often value novelty as contributing to progress. Some ancient authors on scientific subjects also valued new approaches and ideas; as we have seen, the *Aetna* poet emphasized the novelty of his explanation.[12] But other ancient authors valued the idea of working within a tradition, of adding to or refining—if not rejecting—explanations of particular natural phenomena and mathematical problems.[13]

These concerns by authors to engage with the past—whether to innovate in relation to, to participate in, to react against, or even to build upon it—takes us to the question of whether the intellectual (and scientific) community can include people who are no longer living. The evidence strongly suggests that one need not be alive to be an active contributor to—indeed a participant in—the scientific communities peopled by the Greek and Roman authors, their editors, translators, and readers. These authors strongly engage with the past, often citing the work—including the observations, methods, and ideas—of "the ancients."[14] Of course, the commentaries were designed specifically to engage with the ideas of predecessors. And while the texts depend on writing, there was in many cases oral performance and aural engagement with the texts, as demonstrated by the example given by Porphyry, describing the reading out of commentaries to a group of participants gathered together for a lecture.[15]

It is the texts that connect the members of the intellectual community; the participants in the community are the authors, editors, translators, and readers of those texts, who in many cases were divided by vast stretches of time and space. That a sense of membership in a community could transcend historical time and place is shown in Kepler's letter to Galileo dated 13 October 1597, in which he stated: "you are following the lead of Plato and Pythagoras, our true masters."[16] Kepler's engagement with the ideas of the Academy was also demonstrated by his translation of and comments on Plutarch's dialogue.

Throughout this work, I have argued that the genres in which Greco-Roman scientific texts were presented must be seriously considered when we seek to comprehend the information and ideas conveyed. Genre is one of the things that unites authors and readers. Authors, and their readers, bring to texts expectations and shared tacit knowledge regarding the genres of communication, even though genres were created before they were reflected on. A consideration of the genres of texts, and the choices made by authors and their audiences, provides an important way to engage with the intellectual communities in which these texts functioned, providing links between individuals and ideas across the boundaries of historical place, culture, and time.

A Note about Ancient "Books"[1]

The shape of ancient "books" would, no doubt, surprise the modern reader. Furthermore, ancient authors would not recognize the form in which their writings are now available. Little of the physical evidence of ancient scientific books survives. Evidence that survives for the archaic and classical periods is scant; most of our knowledge about the physical characteristics of ancient "books" comes from Hellenistic material. (For our purposes, the death of Aristotle in 322 BCE provides an appropriate date for the beginning of the Hellenistic period.)

The ancients used a wide variety of writing surfaces, including clay tablets, wood, and animal skins, but by far the most widespread material used for writing was papyrus. The plant is native to Egypt and from perhaps 3000 BCE the Egyptians manufactured papyrus sheets to be used for writing. The first-century CE Roman writer Pliny the Elder provided a description (in his *Natural History* 13.74-82) of their production. The sheets were pasted together to form rolls, which seem to have been the standard unit for the manufacture and sale of papyrus, although the widths of rolls varied, as did the length.[2] Because the rolls could not accommodate a large amount of text, in many cases a single work required several papyrus rolls. This is the origin of the division of ancient works into "books."

The papyrus roll was, for much of antiquity, the most common form of "book" and the standard medium for publication. In addition to written descriptions of rolls, such as that contained in Pliny, ancient illustrations of rolls being used by readers exist, including vase paintings from fifth-century BCE Athens.[3] Papyrus rolls, and fragments of rolls, have survived in scattered sites throughout the ancient world.[4]

Publication in the ancient world was, in contrast to much of modern publishing practice before the Internet, extremely casual; no copyrights existed nor were royalties paid.[5] Very little is known about the actual production of texts on papyrus; for example, it is not known whether rolls were copied by dictation, by reading, or both. L. D. Reynolds and N. G. Wilson have emphasized the difficulty of reading ancient texts, which contained rudimentary punctuation and no divisions between words. They suggest that "a high proportion of the most serious corruptions in classical texts go back to [the Hellenistic] period and were already widely current in the books that eventually entered the library of the Museum at Alexandria."[6]

Although the papyrus was the standard form for the publication of literary texts and philosophical treatises, other formats were used, sometimes for specific purposes, such as letters. Between the end of the second and the fourth century CE, the papyrus roll was gradually displaced as the most usual form of book by the codex. The codex format utilized wax-coated wooden tablets and, eventually, parchment leaves. While the precise reasons for the adoption of the codex are not clear, in retrospect scholars have pointed to what may be regarded as its advantages, arguing that the codex was less costly to produce, held more text, and was easier to handle than a roll. Athenian vase-paintings depict readers having trouble with a twisted roll; Pliny the Younger reports that the aged Verginius Rufus broke his hip while trying to recover a dropped roll.[7] C. H. Roberts and T. C. Skeat emphasized the role that the new Christian faith played in the adoption of the codex,[8] but the shift from the papyrus roll to the parchment codex was gradual, involving many innovations and changes in practice. Evidence suggests that Romans, rather than Greeks, first used parchment notebooks. Strikingly, only two Greek writers in the first two centuries CE mention the parchment notebook; one is the physician Galen, who describes a treatment for baldness recorded in a parchment notebook.[9] While the bulk of surviving ancient codices are devoted to Biblical and literary material, they also include astronomical tables, herbals, medical and magical prescriptions, as well as treatises on mathematics, Aristotelian physics, and astronomy.[10]

Many scholars have suggested that it was much easier to find a particular passage in a codex than in a papyrus. The physical difficulty of accessing text is very likely to have affected the ways in which ancient readers, commentators, and editors worked. Some scholars have suggested that the production of standard texts was made easier by the codex, arguing that it is "much harder to make undetectable additions to or deletions from a codex; with a roll, sheets may be pasted in or removed at whim,"[11] but this view is not universally accepted.

The oldest surviving Greek literary texts, the epic Homeric and Hesiodic poems, provide our earliest information about the world of the ancient Greeks and offer a window on a pervasive ancient worldview. The Hesiodic *Works and Days*, a sort of farmers' almanac, contains astronomical lore for predicting the weather; this work may be related to the later *parapēgmata*, which correlated astronomical phenomena with weather.[12]

The earliest figures associated with ancient Greek science, the sixth-century BCE Ionian philosophers Thales of Miletus, Anaximander, and Anaximenes, left no writings that survive. In fact, it is not clear whether Thales wrote anything at all.[13] What is known about their ideas has come through descriptions and quotations in works by later authors. The ancient scholarly tradition of discussing and commenting on the work of predecessors is responsible for what is known about many of the ancient philosophers and mathematicians. In many cases most, if not all, of the surviving information about particular authors and texts is to be found only in writings by a much later author.

We must also consider the survival and transmission of those texts into our own period. We are, in all cases, dependent on versions of the writings of ancient authors that have passed through several filters, or intermediaries, between the author and ourselves. In most cases, any given text will have passed through many hands before it finally reaches a modern reader; the texts will have been copied, translated, and edited, and these activities may each have occurred in any or all of the relevant periods, ancient, medieval, and modern.

Notes

Chapter 1

1. See, for example, Lewenstein (1992) 137. There is an extensive literature on the subject, and Lewenstein provides a useful starting point. Lewenstein also compiled the Cold Fusion Archive, 1989-1993, held in the Division of Rare and Manuscript Collections (Collection number 4451), Cornell University Library.

2. Lewenstein (1992) 140.

3. Sarton (1952/1993) 130. Some scholars favor a later dating for the Homeric poems, but the debate is outside the scope of the present work. Interested readers might turn to Lowenstam (1997) 57-67 for an intriguing consideration of a range of views and evidence.

4. The National Gallery of Art published a catalogue of a joint exhibition with the Metropolitan Museum of Art, by Buitron-Oliver et al. (1992), *The Greek Miracle: Classical Sculpture from the Dawn of Democracy, The Fifth Century B.C.* Some scholars, e.g. Gernet (1983) and Vernant (1982) especially 130-32, (1983) 343-74, have rejected the idea of a Greek miracle. On literacy and other developments, see Burns (1981), referring also to Cornford (1912/1957) and Snell (1953); Harris (1989), particularly chap. 2.

5. See, e.g., Nestle (1942), and the critique in Most (199a). Lloyd (1975/1991) has attacked the view in his "Greek Cosmologies"; see also von Staden (1992). One of the oldest surviving Greek papyri, the Derveni papyrus, is also relevant to the discussion; see, e.g., Betegh (2004).

6. Plato is sometimes offered as an example of a philosopher who rejected traditional mythology, for example in the *Republic* 377a-378d. However, the relationship between mythology and philosophy was complicated, even in Plato's writings; see Morgan (2000); Murray (1999); Rowe (1999); on Aristotle, see Johansen (1999). It should be noted that some ancient philosophers, notably the Stoics, held the view that mythology preserved very ancient wisdom; see Boys-Stones (2001), particularly Part I (Ancient Wisdom: Stoic Exegesis).

7. See Buxton (1999) 8, referring to the work of Vernant (1995) 292-3, 303. Buxton provides a useful overview of some of the issues; see also Most (1999a).

8. I thank Philip van der Eijk for his helpful comments here. As we will see in Chapter 3, Plato referred to the account given in the *Timaeus* as a *mythos*; the main speaker claims to be presenting a "likely story."

9. Calame (1999) 121-22.

10. *Ibid.* 138.

11. *Ibid.* 140.

12. Cf. Plato's *Protagoras*, in which *logos* and *mythos* are said to work together (320b-c, 328c-d).

13. Calame (1999) 140, and in note 34 citing the work of others.

14. *Ibid.* 141, my emphasis.

15. Most (1999b) 334-35.

16. West (1996) notes that "Hesiod's absolute date is now agreed to fall not far before or after 700 BC."

17. Sarton (1952/1993) 160.

18. See, for example, Morgan (2000) 15-45 for a discussion of some of the issues involved in defining "myth" (and *mythos*) in the ancient Greek context. Many of the essays, particularly that by Calame, in Buxton (1999)—including Buxton's own Introduction—usefully demonstrate and discuss the range of meanings attributed to the term *mythos* by modern scholars.

19. While I do not consider medical texts specifically, I recognise that in the relevant historical period the boundaries between medicine and philosophy were very permeable.

20. While some have suggested that the Greek work *epistēmē* is equivalent to the Latin *scientia*, this is open to discussion. Furthermore, *epistēmē*, like the German *Wissenschaft* and French *science*, may be used to describe knowledge of social and historical—as well as natural—phenomena. I am grateful to Philip van der Eijk for comments related to these issues.

21. Aristotle *Metaphysics* 1025b-1026a. The Greek word *mathēma* (plural = *mathēmata*) was used to refer to "that which is learnt," "learning or knowledge," and the "mathematical sciences"; see the entry under *mathē* in LSJ (1940/1968) for examples of usage.

22. For example, Cato the Elder on agriculture, though—as Harry Hine has pointed out to me (private communication)—the contrast between what is Greek and Roman is not as prominent in *On agriculture* as in some fragments of other works, and the biographical tradition about Cato.

23. Duff (2000) xiii. The term "genre fiction" is used to refer to modern works of popular fiction that are regarded as highly standardized, for example historical romances, science fiction, and detective stories. Art historians use the term "genre painting" to refer to a type of painting depicting ordinary activities, rather than historical or mythological subjects.

24. Todorov (1974) 957-58. I am grateful to David Konstan for suggesting I go back to Todorov.

25. Conte and Most (1996) 630-31. See also Conte (1994) chap. 4, "Genre between Empiricism and Theory." Depew and Obbink (2000) "Introduction" provides a useful overview of some of the issues surrounding genres of Greek and Roman literature.

26. Conte and Most (1996) 630-31.

27. Cf. van der Eijk (1997) for a useful discussion of different forms of ancient scientific and medical discourse. Furthermore, concentrating on ideas can obscure our understanding of practice; in the cases of mathematics and technology, see Cuomo (2001, 2007).

28. Lucretius 1: 921-950 and 4:1-25, in almost the same words. On Lucretius, see Kenney (1970).

29. However, the question of the place of poetry within Epicureanism is complicated and tinged with ambiguities; Epicurus himself is sometimes seen as rejecting poetry, while others, including Philodemus, as well as Lucretius, wrote in verse. See the essays in Obbink (1995).

30. I thank Frances Willmoth for this example.

31. Marshall McLuhan (1911-80) argued that "the medium is the message" in his 1964 book, *Understanding Media: The Extensions of Man*.

32. Such work has fruitfully been undertaken by a number of historians of science focusing on the early modern period; see, for example, contributions to Frasca-Spada and Jardine (2000), including Grafton (2000) and Blair (2000). See also Jardine (1991) on the dangers of imposing modern categories on earlier periods.

33. The issue of whether or not the "treatise" was an "actors' category" will be discussed below. Two complex actors' categories that deserve further consideration—and which may be genres—are the demonstration (*epideixis*), which is much discussed by ancient authors, and *historia*, which sometimes refers to an activity, and sometimes to the results of an activity (which may take the form of a specific "genre"). I am grateful to Geoffrey Lloyd for sharing his thoughts on this subject.

34. I recognise that many physicians were also interested in natural-philosophical questions, and that the boundaries between medicine and physics were not always clearly drawn, but leave medical texts for others to consider.

35. Aristophanes (died *c.* 386 BCE) poked fun at philosophers interested in nature in his play *The Clouds*, providing a glimpse of how those engaged in scientific theorizing were regarded by some of their contemporaries.

36. Lloyd's discussion surrounding "Popper versus Kirk: A Controversy in the Interpretation of Greek Science" (1967/1991) is a useful entry point for understanding some of the issues involved.

37. See Lord (1960); Parry (1971); Havelock (1963, 1966, 1971, 1976, 1982, 1986). Cf. Burns (1981) 371 on Havelock's arguing against widespread literacy in Athens during the fifth century BCE; Burns argued that from the end of the sixth century literacy played an important role.

38. Agathemerus 1.1 (= KRS fragment 98, on p. 104); Diogenes Laertius 2.3 (= KRS fragment 138, on p. 143).

39. I thank Geoffrey Lloyd and Philip van der Eijk for their suggestions here.

40. Graf (1996) 1284-85; cf. Burkert (1972) 192-208.

41. On the *akousmata*, see Burkert (1972) 166-92.

42. Diogenes Laertius 8.34-5 = KRS 275 (on p. 230). Diogenes Laertius preserves some of the maxims as *symbola*, the term he uses at 8.17. On the Pythagoreans and beans, see Scarborough (1982).

43. Harris (1989) 66-84, and his chapter 4 more generally.

44. Snyder (2000) 2; see also Konstan (2007a) 3.

45. On libraries, see Platthy (1968). The sources for Pisistratus' establishment of a library (in contrast to his interest in Homer) are rather late; see the testimonia collected in Platthy (1968) 97-110, particularly

numbers 8 (Gellius), 10 (Tertullian), 14 (Hieronymus), 16 (Isidore). Strabo 13.1.54 (C.608) refers to Aristotle's collection of books; cf. Pfeiffer (1968/1978) 67. Harris (1989) 84-86 mentions the book trade and the circulation of written literary texts and (at 85 note 93) the debate regarding Euripides' (probably born in the 480s, died 407-6 BCE) collection of "books" (which would have pre-dated that of Aristotle). See also Small (1997) chapter 3 "The Organization of Collections of Words."

46. See von Fritz and Kapp (1950/1974) 3-7 and Zhmud (2006) 138, who also discusses other lists compiled by Aristotle and his students.

47. See Taub (2003) 93-96.

48. On doxography, see Runia (1999) and Mansfeld and Runia (1997); on Diogenes Laertius, see Warren (2007).

49. On Hippias, see Mansfeld (1986/1990) 27.

50. Runia (1999) 46.

51. See Zhmud (2006) 23-34. (The Greek word for "discovery" or "invention" is *heurēma*; think of what Archimedes was credited with saying in the bathtub: *heurēka* = "I have discovered!".)

52. Nussbaum (1996) 167.

53. Aristotle *Topics* 1.14 (105b12-14), trans. Pickard-Cambridge (*Complete Works*) 1: 175. Here, one can imagine ordered lists, or tables, as Smith (1997) xxiii-xxiv suggests (referring to Aristotle's *Eudemian Ethics* 2,2 1228a28), organized by headings, such as "good" or "animal."

54. Cf. Kahn (1996). On the literary form of the dialogue, see Nightingale (1996) and Blondell (2002); Haslam (1972) and Ausland (1997) treat mimetic aspects of Plato's dialogues.

55. See also Konstan (2007a) 2.

56. Lloyd and Sivin (2002) 129.

57. Barnes (1995) 6-15; quotations from 13-15.

58. Andronicus of Rhodes is credited with having prepared the edition of Aristotle's works which has served as the basis for subsequent editions; his dates are debated, but his editorial work probably took place between 70 and 20 BCE.

59. Aristotle *Complete Works* 1: 74, trans. A.J. Jenkinson.

60. See Plato *Republic* 528d3 and *Theaetetus* 161e6, both cited by Dirlmeier; cf. Dirlmeier (1962) 9-11. Uses of the term *pragmateia* to refer to a "systematic or scientific historical treatise" are found in a number

of authors, including Polybius (second century BCE), Diodorus Siculus (first century BCE), Dionysius of Halicarnassus (first century BCE), who are all considerably later than Aristotle; see the entry in LSJ (1968/1940) for examples. The questions raised by the use of the term are worthy of further study.

61. Plato employed *suggramma* in the *Gorgias* (462b) and *Theaetetus* (166c), while Aristotle uses it in the *Nicomachean Ethics* (1181b2). LSJ (1968/1940) suggest, under the entry for *suggramma*, that *hypomnēma* has an opposite meaning. The entry for *hypomnēma* refers to notes, reminders, or memoranda, but also offers examples of its use to refer to "treatises." I am grateful to David Sider for his suggestions here. See also Fuhrmann (1960) on "systematic" teaching texts.

62. But whatever we make of the "genre" of Aristotle's works, there is an important question regarding what ancient readers made of them. So, for example, we might question the extent to which Aristotle's writings served as a formal model for later writers, but such a study is beyond the scope of this book.

63. The journal article—presented as prose—is an especially important form of modern scientific discourse.

64. Lloyd and Sivin (2002) 129; see also van der Eijk (1997) 92-93.

65. Cf. Cleomedes II.2.7, II.7.12; see the translation by Bowen and Todd (2004) 127 and 165.

66. Nussbaum (1996) 166. See also Barnes (1997) 28-39 on the corpus (versus the "canon") and the role of editors, including Andronicus of Rhodes.

67. Van der Eijk (1997) 89. The Greek term *pragmateiai* may also describe what we regard as "treatises"; see Dirlmeier (1962) 9-11.

68. Cf. Harvey (1937/1974) 349-50, "Prose, 1. Greek Prose."

69. Cf. Pfeiffer (1968/1978) 67.

70. Porphyry *Life of Plotinus* 14.10-14, trans. Armstrong (1966) 41. See also Grafton and Williams (2006) 33-35; Konstan (2007a) 4-5; Snyder (2000) 116-17.

71. Cf. Peters (1967) 22-23.

72. Aristotle *Metaphysics* 982b12-17; trans. Ross, in Aristotle *Complete Works* (1984) 2: 1554.

73. Aristotle *Metaphysics* 995a27-36; trans. Ross, in Aristotle *Complete Works* (1984) 2: 1572-73.

74. Louis (1991) 1: xxiii-xxxv; cf. Inwood (1992). The compiling of a collection of difficulties and problems resonates with other aspects of Aristotle's activities, including the forming of a collection of constitutions, as well as his suggestions for taking reading notes and making lists of opinions.

75. Scholars tend to agree that the author of the so-called Pseudo-Aristotelian *Problems (Problemata)* is not Aristotle, although he is known to have written a book of problems. Some of the material included in the Pseudo-Aristotelian *Problems* seems actually to have its source in the work of Aristotle; several ancient authors (including Plutarch and Cicero) described portions of the *Problems* as Aristotelian. See Hett (1957) vii. Cf. also Louis (1991).

76. Hine (1981) 28. See also Cherniss (1976/2000) 2-5, in which he discusses the *zētēmata* literature, which posed questions concerned with the meaning of a passage in a text (traditionally in Homer, but also applied to other texts as well).

77. On the *Problems* see Flashar (1975) and Sharples (2006), as well as the other articles in De Leemans and Goyens (2006).

78. In the *Posterior Analytics* (B 1.89b24-35), Aristotle had explained that there are four main types of question to be posed in an investigation: the "that," the "why," the "if it exists," and the "what it is." The form of the questions posed in the *Problems* is an enquiry of the second type. See Mansfeld (1992) 70-71.

79. I thank Philip van der Eijk for this point.

80. Pseudo-Aristotle *Problems* 939a110-15, trans. Forster (*Complete Works*) 2:1471, question 9.

81. Pseudo-Aristotle *Problems* 912b4-10, trans. Forster (*Complete Works*) 2:1419, question 10.

82. However, it is not clear that this problem is about mathematical theory; rather, the question concerns the shadows cast by the sun and moon. Here, a geometrical demonstration is used to present an argument about phenomena.

83. Taub (2003) 103; Netz (1999) 37 note 65. See also Sider (2005) 15-19 on diagrams in ancient texts.

84. On Eratosthenes' *Letter to King Ptolemy*, see Taub (2008).

85. See Trapp (1996).

86. On a letter attributed to Eratosthenes, directed to his royal patron, and announcing a solution to a geometrical problem, see Taub (2008). During the early modern period, dedicatory letters were often used to advertise patronage.

87. Diogenes Laertius 2:613-15, trans. Hicks. My emphasis.

88. Geminus 1.1-2, trans. Evans and Berggren (2006) 113.

89. Nicomachus 1.1, trans. D'Ooge (1991) 599.

90. See Toomer (1996b) for publication details for these commentaries on Nicomachus. More generally, see Mansfeld (1994) Chapter 1 ("*Schemata Isagogica* from Origen to Stephanus, and some precedents").

91. Aristotle *Topics* 105b15-17, trans. Pickard-Cambridge (*Complete Works*) 1:175.

92. Simplicius *On Aristotle's* Physics 2, trans. Fleet (1997) 45, slightly amended. Instead of "physicist," Fleet translates "natural scientist"; Hardie and Gaye in their translation of Aristotle's *Physics* (in *Complete Works* 1:331) use the phrase "student of nature."

93. See von Staden (2002); Sluiter (2000).

94. Quoted by Grafton and Williams (2006) 34, citing Porphyry *Life of Plotinus* 14.10-14. On commentaries more generally, see Most (1999c) and Gibson and Kraus (2002).

95. On higher education in late antiquity, see Cameron (1967).

Chapter 2

1. See chapter 1 for questions regarding the authorship and dating of the Homeric poems.

2. There are many shorter poems that deal with nature, such as those collected in the Greek anthologies; see, for example, the *Greek Anthology*, trans. W. R. Paton, 5 vols., Cambridge, MA: Harvard University Press, *LCL*, 1916-19; "Arithmetical problems, riddles and oracles" (Book 14 of the *Anthology*) can be found in volume 5: 25-107.

3. For example, Archimedes' 'Cattle problem,' in Thomas (1941), 2: 202-5.

4. Plutarch's dialogue *On the Face on the Moon*, which is discussed in the following chapter, includes quotations from a number of poets.

5. In *On Agriculture* Columella wrote on gardening in verse (in Book Ten) and in prose (in Book Eleven).

6. Paisley and Oldroyd (1979) 1, quoting from 10.

7. Here, and in what follows, the translation—presented as prose, rather than verse—is that of Duff and Duff (*LCL*). Although *Aetna* was composed as a poem in Latin, to my knowledge no English verse translation of the work has been published; consequently, readers of the work in English need to be reminded that it was originally presented as a poem. In the section 'From Poetry to Philosophy?' below, some features of ancient poetry are discussed.

8. Most (1999b) 332-33, citing Aristotle *Poetics* 1447b17-20, referring to Homer and Empedocles; the passage is quoted below.

9. Guthrie (1962) 1: 450 n. 2 thought that neither Plato nor Aristotle "can be supposed to have been very serious in this." Cf. Ross (1924) 1: 130 note 27 on *Metaphysics A* 983b27f., who remarked that Plato "jestingly suggests that Heraclitus and his predecessors derived their philosophy from Homer, Hesiod and Orpheus," and added that Aristotle admits that "the suggestion has no great historical value."

10. See Lamberton and Keaney (1992), especially Lamberton's "Introduction," particularly xvi. See also Richardson (1992) and Long (1992). Konstan (1991) 21 suggested that, in concentrating on the mythic genealogy of the gods in the *Theogony*, Hesiod left room, so to speak, for materialist physical explanations of the world.

11. Porphyry in Schrader (1880) 1.240.14 = DK 8A2; cited by Konstan in Russell and Konstan (2005) xiii-xiv. On allegory more generally, see Konstan's "Introduction" in Russell and Konstan (2005), and also Ramelli and Lucchetta (2004).

12. See, for example, Lloyd (1975/1991) "Greek Cosmologies" on the transition from mythology to philosophical cosmology, and on the dangers of over-simplification.

13. See, for example, Plato *Timaeus* 39e-41a; Epicurus *Letter to Pythocles* 97; Lucretius 1.44-49, 2.646-51, 5.146-55. Cf. Taub (1993) chap. 5; Taub (forthcoming).

14. On Hesiod, see Pliny *Natural History* 18.201; cf. 18.212-13.

15. Parmenides' writings survive only as fragments, which were preserved largely by Sextus Empiricus (end of the second century CE) and Simplicius (sixth-century CE commentator who transcribed

extracts of Parmenides' poem into his commentaries on Aristotle's works *On the heavens* [on cosmology and astronomy] and *Physics*); Simplicius explained that his motive in copying extracts of the poem was to preserve it, because of its scarcity. See Schofield in KRS (1983) 241.

16. Meter varies with the arrangement of groups of "long" and "short" syllables, defined by the length of time taken to pronounce. A "foot" consists of a group of syllables; the foot known as the "dactyl" (literally, "finger") consists of three syllables: one long, followed by two short. Dactylic hexameter consists of six dactylic feet; sometimes a "spondee" (composed of two long syllables) is substituted for a dactyl. For a more detailed discussion, see, e.g., Harvey (1974) 268-72.

As an example in English, hexameter was used by the American poet Henry Wadsworth Longfellow in his poem *Evangeline* (first published in 1841). Here is the opening line:

This is the forest primeval. The murmuring pines and the hemlocks. . . .

The pattern is long-short-short, repeated for five feet; the final foot is long-short (a trochee). Cf. Campbell (1908).

17. Schofield in KRS (1983) 241.

18. Trans. KRS (1983) 243.

19. The etymology of the English term is not entirely clear; compare, for example, *The Compact Edition of the Oxford English Dictionary* (Oxford: Oxford University Press, 1971) and *The American Heritage Dictionary of the English Language* (Boston: American Heritage Publishing, 1973).

20. Mourelatos (1970) 45-46.

21. Schofield in KRS (1983) 283; only fragments of the poems survive.

22. Aristotle *Poetics* 1447b17-20.

23. Plutarch *Quaestionum convivalium libri ix* (*Table-talk*) 683E, trans. KRS (1983) 284 note. The questions, or problems, posed are presented as dinner-table topics for Plutarch, his family, and friends. The first question posed is whether philosophy is a suitable topic for conversation at a drinking party.

24. Schofield in KRS (1983) 283.

25. Aristotle *On the generation of animals (De generatione animalium)* A 23, 731a4 (=KRS (1983) fragment 385, on p. 306).

26. Cf. Most (1999b). I am grateful to David Konstan for his comments here.

27. Strabo 1.2.3 (C.15-16), trans. Jones, 1: 55.
28. Lucretius 1: 921-950 and 4:1-25, in almost the same words.
29. Toomer (1996a); see also Kidd (1997) Introduction.
30. Volk (2002) 34-43 discusses didactic poetry as a genre. She also referred to two late ancient authors who considered it to be a separate genre; Diomedes (4th/5th century CE), the author of an *Ars grammatica*, and the anonymous Peripatetic author of the *Tractatus Coislinianus*. However, it is not clear whether didactic was recognised earlier as a genre. While some scholars, including Schiesaro (1996), have expressed caution about viewing "didactic poetry" as a genre, for the purposes here I will (cautiously) follow Volk's lead, and take a somewhat "empirical" approach. Those poems that are primarily regarded as didactic typically also display philosophical, political, and religious, as well as literary, concerns. See also the various articles in Schiesaro, Mitsis, and Clay (1993).
31. Toohey (1996) 3.
32. Thalmann (1984) 120-21, and in his Introduction, emphasized that hymns should be understood as part of the broader category of epic poetry; the Greek term *ta epea* referred to poems in dactylic hexameter.
33. Parker (1996). See also Hunter (1996) 46-52 on hymns.
34. The range of scholarship on Lucretius is vast. Interested readers are directed, as a beginning, to Fowler and Fowler (1997/1999) xi-xxxii, as well as their "Select Bibliography," xxxiii-xxxvi.
35. Trans. Melville (1997/1999) 3.
36. Trans. Melville (1997/1999) 54. In some editions, these lines are printed in Book 1.44-49 as well, but were arguably wrongly inserted there at some point in the tradition; I am grateful to Philip Hardie for his comments here.
37. Furley (1970/1989) pointed to similarities between Lucretius' proem and some ideas associated with Empedocles regarding nature (particularly the four elements), suggesting that Lucretius was influenced in his philosophy by Empedocles. See also Clay (1983) *passim*. Sedley (1998) 1-34 has argued that, rather than adopting Empedocles' natural philosophical content for his poem, Lucretius borrowed the form. The proem of Empedocles poem *On nature* does not survive, but Sedley (1998) 26 has argued that it is entirely plausible that his proem would have opened with a hymn to Aphrodite.

38. Sedley (1998) 16.

39. Trans. Melville (1997/1999) 54.

40. Scholars debate the question of whether or not Epicurus believed that gods are really existent; see, for example, Obbink (1996) and Long and Sedley (1987) 1: 139-49.

41. Sedley (1998) 27.

42. On Epicurean theology, see Long and Sedley (1987) 1: 63-4, 139-49, and also Mansfeld (1999) esp. 454ff., 463f., 478.

43. See, for example, Cicero *On the Nature of the Gods* 2.22-3, 28-30; Diogenes Laertius 7.137-39. See also Sambursky (1959) 21-44, particularly pp. 36-37. Our knowledge of the natural philosophy of the early Stoics, including Zeno (of Citium in Cyprus, 335-263 BCE), the founder of the school, and his successors Cleanthes (of Assus in Asia Minor, 331-232 BCE) and Chrysippus (of Soli in Cyprus, *c*. 280-207 BCE), is based on reports preserved in later writers. On the Stoics, see Long and Sedley (1987) 1:266-343; Hahm (1977).

44. Nothing is known about Manilius' life. Goold (1977) xii dates Books 1 and 2 to the period when Augustus (who died in 14 CE) was still alive; Books 4 and 5 were written after his death.

45. Toohey (1996) 180-84 argued that Manilius was writing in direct opposition to Lucretius.

46. Trans. Goold (1977) 5-7.

47. Cf. Volk (2002) 234.

48. However, the strength of this argument is not entirely clear. On the dating of the poem, see Goodyear (1965) 56-59, who is responsible for the standard text of the poem, and Goodyear (1984). On this question, as well as others relating to the poem, see Volk (2005).

49. I am grateful to Harry Hine for helping me to clarify this point.

50. Seneca *Epistle* 79.5; cf. Goodyear (1984) 350-51. Hine (1981) 32-34 considered possible reasons for the absence of a discussion of volcanoes in Seneca's *Natural Questions*.

51. Here and in what follows, the translations of *Aetna*, unless otherwise noted, are by Duff and Duff, in the Loeb edition; here, pp. 411-15.

52. Hadrian's ascent is described (in the much later *Scriptores Historiae Augustae* Hadrian 13.2, in *LCL* 1: 40-41), and Seneca suggested to his friend Lucilius that he might climb Etna (*Epistle* 79.2). See Hine (2002) 57, note 9.

53. Lucretius speaks about the difficulties of his own task at 1.136-45.

54. I have substituted "Tartarus," as it appears in the Latin, for the translation "hell" suggested by Duff and Duff. The text of the poem is corrupt in places, and different systems are used to number the lines in this part; the line numbers of the Loeb edition are given here.

55. Epicurus *Letter to Herodotus* 79-80.

56. This is not the usual picture that we have of Romans engaged in scientific pursuits; typically, historians have emphasised what they regarded as a Roman preoccupation with the practical. Cf. Lindberg (1992) 137; see also, for example, Reymond (1927) 92. Cicero (*Tusculan Disputations* 1.2) praised his fellow Romans for limiting their study of mathematics to what was necessary for the utilitarian purposes of measuring and calculating, rather than pursuing geometry in the manner undertaken by the Greeks.

57. Woodhead and Wilson (1996).

58. Cf. Hine (1996), (2002). A more literal translation is "mill-stone."

59. So, for example, on wind as a possible cause of thunder, see Theophrastus in Daiber (1992) 261-62, where four of seven possible causes of thunder are linked to wind; Epicurus *Letter to Pythocles* 100 and Lucretius 6.121-31; on wind as a possible cause of earthquakes, see Epicurus *Letter to Pythocles* 105.

60. At 172-74 the *Aetna* poet refers to evidence to support the view that the universe will eventually "return to its primeval appearance." Epicurus and Lucretius both used an analogy of a living organism, with a fixed life-cycle, to defend some of their ideas about the world; see, for example, Lucretius 2.1122-74, and cf. Furley (1999) 425. Asmis (1984) 314-15 comments briefly on the idea that the process of growth and decline of worlds is analogous to the growth and decline of living beings. See also Solmsen (1953) and Taub (forthcoming). I am grateful to David Sedley for his comments; on the use of analogy more generally, see Lloyd (1966).

The poet Marcus Manilius (early 1st century CE) describes the world (*mundus*) as a living creature (for example, at 2.63-66); reports of related Stoic views can be found in *Stoicorum Veterum Fragmenta* 2.633-645. See also Volk (2005) 81-82.

61. Volk (2002) 247.

62. On *topos*, see Most and Conte (1996).

63. Trans. J.E. Sandys, modified by Hine (2002) 70.

64. Seneca *Epistle* 79.5.

65. Goodyear (1984) 346-47.

66. I am grateful to Philip Hardie for his suggestions here.

67. Solmsen (1957), citing Sudhaus (1898) [2-3].

68. Lucretius and Virgil both wrote about nature, but had different philosophical and moral interests.

69. On the brothers' act of filial piety, cf. Strabo 6.2.3 (C. 269), Seneca *On Benefits* (*De Beneficiis*) 3.37.2 (*LCL*, 3: 200-01), Martial 7.24.5 (*LCL*, *Epigrams* 2: 94-95). Claudian *Shorter Poems* (*Carmina Minora*) 17 (L) (*LCL*, 2:188-91) contains a poem *On the Statues of Two Brothers at Catina* (*De piis fratribus*).

70. Toohey (1996) 191-92 addressed this briefly.

71. Toohey (1996) 190-91 also noticed the tension in the poem.

Chapter 3

1. Erasmus Darwin's *The Botanic Garden* (first published 1789-91, and reprinted repeatedly) is an example of a late eighteenth-century poem on a "scientific" topic.

2. For the translation see Kepler (1870) 8: 76-123. On Plutarch's influence on Kepler, see Dick (1980) 9 and Dick (1982) 22, 70-71, 75; Grafton (1997) 208-13.

3. I thank Philip van der Eijk for raising this point with me.

4. Cf. Kirk, in KRS (1983) 74; Lloyd (1987) 2, 50-108, *et passim*.

5. At various points in the *Republic* Plato criticizes myth. See, for example, Plato *Republic* 377a-378d, where it is argued that conventional mythology is dangerous, and 604b-608a, in which the value of *logos* (argument) over poetry is stressed. Cf. Morgan (2000) 207-08; Most (1999b) 339.

6. The examples given here are to be found in LSJ (1940/1968), under the entry for *dialogos*.

7. Together with Plato's contributions, those of Plutarch represent what survives of Greek philosophical dialogues. While Lucian (born *c*. 120 CE) wrote dialogues in Greek, he is generally considered to be more of a satirist or belletrist than a philosopher. Those by Aristotle do not

survive, except for some fragments. In Latin, several of Cicero's dialogues deal with natural-philosophical topics; Varro's (116-27 BCE) *De re rustica*, on farming, is cast as a dialogue.

8. Rowe (1996) 462.

9. Lloyd and Sivin (2002) 129.

10. For example, in Book 7 of the *Republic* Glaucon and Socrates discuss the usefulness of mathematical subjects, including astronomy, as preliminary studies prior to undertaking dialectic; Plato's interests were very broad, but mathematics had a special place in the educational system he promoted. See Burnyeat (2000).

11. See, for example, Plutarch's *Table Talk* (*Moralia* 8 & 9, *LCL*), and Athenaeus' *Deipnosophistai* (*The Learned Banqueters* or *The Sophists at Dinner*), as well as Plato's *Symposium*.

12. See Sedley (2007) chap. 4, on the *Timaeus-Critias* dialogue, particularly 95-97.

13. Lamberton (2001) 147.

14. *Ibid.* 172.

15. *Ibid.* 26.

16. Several ancient authors offered what might be regarded as satirical or even pseudo-scientific narrative accounts of travel to the moon; see, for example, Lucian's *A True Story* and Antonius Diogenes' *The Wonders Beyond Thule* (described in Photius' Summary (*Bibliotheca* 166), produced in the ninth century CE); see Reardon (1989) for translations of both.

17. Both manuscripts are in the Bibliothèque Nationale, Paris, one dating from the fourteenth, the other from the fifteenth century; their relationship is not entirely clear, but there are strong arguments supporting the view that one is descended from the other. See Cherniss (1957/1984) 26-7; on the question of mutilation, Cherniss (1957/1984) 2.

As it stands, the structure of the work does not unfold until the reader is well into the text; Cherniss (1957/1984) 14 provides a summary of the structure.

18. Strangers play important roles in some Platonic dialogues, notably the *Statesman* and the *Laws*.

19. Trans. Cherniss (1957/1984) 35. All translations from Plutarch's dialogue are by Cherniss, unless otherwise noted.

20. See Cherniss (1957/1984) 15 for a detailed discussion of this point.

21. Cherniss (1957/1984) 16-17.

22. Trans. Cherniss (1957/1984) 41; the words in angle brackets have been supplied by the editor. On ancient mirrors, see Lloyd-Morgan (1996).

23. Trans. Cherniss (1957/1984) 107-9. The description of the mirrors is somewhat difficult to follow. It is possible that Plutarch intended the passage to be a bit opaque (cf. his comments in *On Listening to Lectures* (*De audiendo*) 43a-b (*LCL* 1:230-33) on the use of technical terms, mentioned below). Cherniss (1957/1984) 109, note a, has suggested that some of the difficulty may be due to Plutarch's misunderstanding a passage in Plato's *Timaeus* (46b-c).

24. Posidonius is mentioned at 929D, Aristarchus at 932B.

25. I have not been able to consult the manuscripts, but there is no indication in Cherniss' edition that they contain diagrams.

26. Lamprias asks Theon to identify a quotation, from Sophocles, at 923F; cf. 931E, 940A for references to his literary knowledge.

27. Plutarch *On the Face on the Moon* 937D.

28. The eschatological myth is preceded by a short geographical introduction (941A-942C), referred to briefly below.

29. Cf. Cherniss (1957/1984) 18.

30. The stranger is mentioned for the first time, in the extant text, at 942B. The geographical introduction (941A-942C) has itself attracted a great deal of interest; see, for example, Coones (1983).

31. Kepler (1870) 8: 119-20, notes 97 and 98; cf. Cherniss (1957/1984) 21.

32. Cherniss (1957/1984) 14.

33. Cherniss (1957/1984) discusses the participants at 3-9 and hints at the ambiguities at 15-16.

34. Lamberton (2001) 174.

35. Some of the participants here play a role in other of Plutarch's dialogues, and there may be the expectation that readers will be familiar with these. For example, Plutarch identifies Lamprias as his brother in, for example, *The E at Delphi* (at 385D), in *Moralia* 5: 204-5 (*LCL*); Plutarch himself does not appear in the extant portions of *On the Face on the Moon*.

36. Cf. Cherniss (1957/1984) 15-16 who favors the view that Pharnaces was not present for the earlier conversation, but also offers evidence that he was; by Lamberton's reading (2001) 174, Pharnaces was present.

37. Carthage, on the coast of what is today Tunisia, was a Phoenician colony and later a major Roman city, with very impressive building works; during the second century CE it had the longest aqueduct in the Roman world.

38. Lamberton (2001) 150. See also Sedley (2007) 54-57 and 128-32.

39. In the *Critias* dialogue the conversation is resumed, but breaks off in the middle of Critias' account. The extended discourse of Timaeus can be understood as a variant or sub-genre of dialogue; other ancient dialogues, notably some of Cicero's, also contain lengthy speeches or monologues. Aristotle's dialogues also may have had lengthy speeches. Cf. Rowe (1996), and Sedley (2007) 95-97, who cites the work of Osborne (1996), Broadie (2001) and Johansen (2004) on the dialogic context of Timaeus' speech. Johansen's chap. 9, "Dialogue and dialectic," is particularly helpful on the topic.

40. Trans. Cornford (1937/1975) 20.

41. Trans. Cornford (1937/1975) 23.

42. See Rowe (1999).

43. Sedley (2007) 99-100. See also Most (1999a) 34.

44. Sedley (2007) 97. As Sedley and Cornford (1937/1975) 31 have pointed out, even though the *Timaeus* is written as prose, it follows many of the conventions of poetry.

45. Taylor (1928) 59.

46. Sedley (2007) 97.

47. Trans. Cornford (1937/1975) 73; see also his commentary, p. 76.

48. Cornford (1937/1975) 74-76. Sedley (1976) has pointed to the literary evidence of the school of mathematicians at Cyzicus, followers of Eudoxus (fourth century BCE), who had a special interest in building instruments to demonstrate his ideas of planetary motion.

49. Cherniss (1957/1984) 18-19.

50. Cherniss (1957/1984) 35 pointed to passages in Plato *Phaedo* 77e and 114d, and *Republic* 608a, which refer to the chanting of incantations.

51. Plato uses myths in some of his other dialogues. I am grateful to David Konstan for reminding me that in the *Phaedo* Plato's approach may be more similar to that which we later encounter in Plutarch's *On the Face on the Moon*. On the myth that closes the *Phaedo*, see Sedley (1990).

52. Lamberton (2001) 26.

53. Gadamer (2004) 371.

54. Letter-writing can, to some extent, also be considered a form of conversation.

55. Gadamer (2004) 360.

56. *Ibid.* 340.

57. See, for example, Galilei, *Dialogue concerning the Two Chief World Systems*, trans. Drake (1967).

58. Plutarch *On the Festival of Images at Plataea*, in *Moralia* 15 (*Fragments*): 284-85 (*LCL*), trans. Sandbach, slightly amended. Sandbach has noted (282-83) that some scholars have suggested that this fragment comes from a dialogue by Plutarch, and does not represent his own views.

59. Johansen (2004) 177-79.

Epilogue

1. There is a vast literature addressing related issues.

2. I am grateful to Harry Hine for raising these questions with me.

3. See chapters 1 and 2.

4. I have written on other genres of scientific communication used by the ancient Greeks and Romans; see Taub (2007) on biography and Taub (2008) on a "mathematical" letter. I am currently writing a book that will consider a broader range of Greek and Roman scientific and mathematical genres.

5. Konstan (2007b). See also Cavallo (1999) 88, on the practice of writing in a book, afforded by the adoption of the codex.

A number of intriguing issues might be raised, but cannot be considered in any detail here. For example, in Plato's *Phaedrus* (274b ff.), the idea that writing can convey knowledge is challenged; but contrast this concern about the deficiencies of writing with the way in which Plato's ideas are encountered: through reading (and discussing) his own written works, the dialogues. There are also questions relating to what readers engage with when they confront a written work: the text or the author? To what extent is a text a "closed" work? Readers in various communities may treat written works differently; for example, literary scholars or historians may have rather different views than philosophers about what is being encountered in a text.

6. For example, the Roman translators of Aratus, including Cicero, Germanicus, and Avienus.

7. See Frasca-Spada and Jardine (2000) Introduction, especially pp. 2 and 4. On genres as shared systems of expectations, see Hirsch (1967) chapter 3.

8. Konstan (2007b) 2-3. On Plutarch as an active reader, who demands similar activity from other readers, see also Konstan (2004).

9. Johnson (2000) 623. Active reading practices were by no means restricted to the ancient period; ancient scientific poetry and dialogues were closely read, edited, commented upon, and translated during the early modern period, by "scientific" authors such as Kepler, who, as we have seen, translated Plutarch's dialogue into Latin. However, the active reading, collecting, editing, and translating practices of the medieval and early modern periods cannot be dealt with here.

10. For example, see Simplicius *De caelo* 557, 25ff.=KRS 288 (on pp. 242-43); see KRS (1983) 244-62 for other portions of Parmenides' poem preserved by Simplicius.

11. I thank Michael Worton for highlighting this distinction.

12. But this emphasis on innovation may have been a convention of Latin didactic poetry; I thank Philip Hardie for this suggestion. On some ways of thinking about the past in antiquity, see Edelstein (1967), Boys-Stones (2001), Zhmud (2006).

13. See Taub (2003) 187-89.

14. See, for example, Ptolemy's reference to "the ancients" in the *Almagest* 1.3; trans. Toomer (1984) 38.

15. See Grafton and Williams (2006) 33-34, citing Porphyry *Life of Plotinus* 14.10-14; also cited by Konstan (2007a) 2-4. See also Snyder (2000) 116-17.

16. Kepler (1951) 40-41; cf. (1945) 145. See Taub (2007). See Jardine and Segonds (1999) 222ff. with regard to Kepler's interest in Pythagoreanism.

A Note about Ancient "Books"

1. Much of the information here appeared in Taub (2000) 27-29 and is reprinted (with minor changes) with kind permission from Andrew Hunter, editor of the volume *Scientific Books, Libraries and Collectors*, 4th edn (Aldershot, England: Ashgate).

2. For details on the production of papyrus, ink, and books, see Stephens (1981) 421-36. She has pointed out that Pliny's description is not always accurate. See also Reynolds and Wilson (1991) 1-5, and Small (1997) chapter 2 "Ancient Books."

On the history of papyrus, see Lewis (1974). The standard introductory work in English on papyrology is Turner (1968/1980). See Bagnall (1995) for a discussion of historiographical and methodological issues. The bibliography in Rupprecht (1994) is particularly useful.

3. A vase painting of Sappho is mentioned by Stephens (1981) 426; also see Plate 8 Lewis (1974).

4. Over eighteen hundred rolls (most irretrievably charred), the remains of a library, were found at Herculaneum, a city buried following the eruption of Vesuvius in 79 CE; these are particularly useful for historians of ancient philosophy.

5. Reynolds and Wilson (1991) 24. As Reynolds and Wilson explain, authors could make changes to a text by asking friends to alter their copies, but other copies would remain unchanged. See also Small (1997) chapter 3 "'Publication'."

6. Reynolds and Wilson (1991) 4-5.

7. Turner (1968/1980) 7, citing Beazley (1948) and Immerwahr (1964). On the accident, see Pliny *Letters* Book 2. 1. 5 (To Vocanius Romanus), trans. Radice (*LCL*) 1:79-81.

8. Colin H. Roberts, in a study originally published in 1954 and revised with T. C. Skeat in 1983 as *The Birth of the Codex*, argued that the preference for the codex was linked to the rise and spread of Christianity.

9. Galen, *Opera*, ed. Kuhn, xii. 423, cited in Roberts and Skeat (1983) 22.

10. Turner (1977) discussed the form, manufacture, and contents of early codices.

11. Stephens (1981) 433. See also Reynolds and Wilson (1991) 34-35.

12. See Taub (2003) chapters 1 and 2.

13. According to Simplicius (*Commentary on Aristotle's Physics*, p. 23, 29, ed. Diels=KRS 81 [on p. 86]), Thales left no writings other than "the so-called 'Nautical Star-guide'." Diogenes Laertius (Book 1. 23=KRS 82 [on pp. 86-87]) claimed that the starguide was attributed to Phokos of Samos.

Bibliography

The abbreviation *LCL* indicates a volume in the *Loeb Classical Library*.

Ancient texts

Aetna, trans. R. Ellis, Oxford, England: Clarendon Press, 1901.
——— in *Minor Latin Poets*, trans. J. Wight Duff and Arnold W. Duff. London: William Heinemann, and Cambridge, MA: Harvard University Press (*LCL*), rev. edn 1935, 358-419.
Antonius Diogenes *The Wonders Beyond Thule*, summarized by Photius in *Biblioteca* 166, in Reardon, B. P. (ed.) *Collected Ancient Greek Novels*, Berkeley: University of California Press, 1989, 775-82.
Archimedes "Cattle problem," in *Greek Mathematical Works*, trans. Ivor Thomas, 2 vols, Cambridge, MA: Harvard University Press (*LCL*) 1941, 2: 202-5.
[Aristotle] *The Complete Works of Aristotle, The Revised Oxford Translation*, ed. J. Barnes, 2 vols, Princeton, NJ: Princeton University Press, 1984, Bollingen Series 71.2.
——— *Physics,* trans. R. P. Hardie and R. K. Gaye, in *The Complete Works of Aristotle,* vol. 1.
——— *Problems,* trans. E. S. Forster, in *The Complete Works of Aristotle*, vol. 2.
——— *Problems*, trans. W. S. Hett, 2 vols, London: Heinemann (*LCL*), 1936-37.
——— *Topics*, trans. W. A. Pickard-Cambridge, in *The Complete Works of Aristotle*, vol. 1.
Athenaeus *The Deipnosophists*, trans. Charles Burton Gulick, 7 vols, London: Heinemann (*LCL*), 1927-41.
Cato *On Agriculture (De re rustica)*, trans. W. D. Hooper, rev. H. B. Ash, in *Marcus Porcius Cato: On Agriculture; Marcus Terentius Varro: On Agriculture*, Cambridge, MA: Harvard University Press (*LCL*), 1935.
Cicero *Letters to Atticus*, trans. E. O. Winstedt, 3 vols, Cambridge, MA: Harvard University Press (*LCL*), 1912-18.
——— *On the Nature of the Gods (de natura deorum)***,** trans. H. Rackham, in *Cicero, De Natura Deorum, Academica*, Cambridge, MA: Harvard University Press (*LCL*), 1933.

———— *Tusculan Disputations*, trans. J. E. King, London: Heinemann (*LCL*), 1927.
[Claudian] *Claudian* trans. M. Platnauer, 2 vols, Cambridge, MA: Harvard University Press (*LCL*), 1922, vol. 2 (*Shorter Poems*).
[Cleomedes] *Cleomedes' Lectures on Astronomy: A Translation of the Heavens,* with an Introduction and Commentary by Alan C. Bowen and Robert B. Todd, Berkeley: University of California Press, 2004.
Collected Ancient Greek Novels, Reardon, B.P. (ed.), Berkeley: University of California Press, 1989.
Columella *On Agriculture (de re rustica),* trans. H. B. Ash (Books 1-4), E. S. Forster and E. H. Heffner (Books 5-12) in *Lucius Junius Moderatus Columella: On Agriculture,* ed. and trans. E. S. Forster and Edward H. Heffner, 3 vols, Cambridge, MA: Harvard University Press (*LCL*), 1941-55.
Diogenes Laertius *Lives of Eminent Philosophers,* trans. R. D. Hicks, 2 vols., Cambridge, MA: Harvard University Press (*LCL*), 1925.
Epicurus *Letter to Herodotus,* in *Diogenes Laertius: Lives of Eminent Philosophers* 10. 34-83, trans. R. D. Hicks, 2 vols, Cambridge, MA: Harvard University Press (*LCL*), 1925, 2: 564-613.
———— *Letter to Pythocles,* in *Diogenes Laertius: Lives of Eminent Philosophers* 10. 93-117, trans. R. D. Hicks, 2 vols, Cambridge, MA: Harvard University Press (*LCL*), 1925, 2: 612-43.
[Geminus] *Geminos' Introduction to the Phenomena: A Translation and Study of a Hellenistic Survey of Astronomy,* trans. and commentary James Evans and J. Lennart Berggren, Princeton, NJ: Princeton University Press, 2006.
Greek Anthology, trans. W.R. Paton, 5 vols., Cambridge, MA: Harvard University Press, Loeb Classical Library, 1916-19.
[Hesiod] *Hesiod: Homeric Hymns and Homerica,* trans. H. G. Evelyn-White, , Cambridge, MA: Harvard University Press (*LCL*), 1914.
———— *Hesiod: The Poems and Fragments,* trans. A. W. Mair, Oxford, England: Clarendon Press, 1908.
———— *Hesiod: The Works and Days, Theogony, The Shield of Herakles,* trans. R. Lattimore, Ann Arbor: University of Michigan Press, 1959.
———— *Hesiod: Works and Days, Theogony,* trans. S. Lombardo, Indianapolis, IN: Hackett, 1993.
Homer *Iliad,* trans. R. Lattimore, *The Iliad of Homer,* Chicago, IL: University of Chicago Press, 1951, rpt. 1961.
———— *Odyssey,* trans. R. Lattimore, *The Odyssey of Homer,* New York: Harper & Row, 1965, rpt. 1975.

Lucian *A True Story*, in Reardon, B. P. (ed.) *Collected Ancient Greek Novels*, Berkeley: University of California Press, 1989, 619-49.

Lucretius *On the Nature of the Universe* (*De rerum natura*), trans. R. Melville, Oxford, England: Oxford University Press, Oxford World's Classics, 1997/1999; first published 1997 by Clarendon Press.

—— *De rerum natura*, trans. W. H. D. Rouse, Cambridge, MA: Harvard University Press, rev. 2nd edn, 1982.

Manilius *Astronomica,* trans. G. P. Goold, Cambridge, MA: Harvard University Press (*LCL*), 1977.

Martial *Epigrams,* trans. D. R. Shackleton Bailey, Cambridge, MA: Harvard University Press, 1993.

Nicomachus *Introduction to Arithmetic*, trans. Martin L. D'Ooge, in *Great Books of the Western World*, vol. 10, *The Thirteen Books of Euclid's Elements, The Works of Archimedes including the Method, Introduction to Arithmetic by Nicomachus*, Chicago, IL: Encyclopedia Britannica, 1991.

Pindar *The Odes of Pindar: including the principal fragments*, trans. John Sandys, London: Heinemann (*LCL*), rev. edn. 1937.

—— *Olympian Odes; Pythian Odes*, trans. William H. Race, Cambridge, MA: Harvard University Press (*LCL*), 1997.

Plato *Timaeus* in *Plato's Cosmology*, trans. Francis MacDonald Cornford, Indianapolis, IN: Bobbs-Merrill, 1937, rpt. 1975.

—— *Timaeus*, trans. B. Jowett, in *Plato: The Collected Dialogues*, E. Hamilton and H. Cairns (eds), Princeton, NJ: Princeton University Press, Bollingen Series 71, 1961, rpt. from *The Dialogues of Plato*, 4th edn, 1953; 1st edn 1871.

—— *Plato: Complete Works*, ed. J. M. Cooper, D. S. Hutchinson, J. Barnes, Indianapolis, IN: Hackett, 1997.

Pliny the Elder *Natural History*, trans. H. Rackham *et al.*, *Pliny: Natural History*, 10 vols, Cambridge, MA: Harvard University Press (*LCL*), 1938-63.

Pliny the Younger *Letters and Panegyricus*, trans. Betty Radice, 2 vols, Cambridge, MA: Harvard University Press (*LCL*). 1969.

Plutarch *The E at Delphi*, in *Plutarch's Moralia* 5, trans. Frank Cole Babbitt, Cambridge, MA: Harvard University Press (*LCL*), 1936.

—— *The Obsolescence of Oracles* in *Plutarch's Moralia* 5, trans. Frank Cole Babbitt, Cambridge, MA: Harvard University Press (*LCL*), 1936. [Also referred to as *The Disappearance of Oracles*.]

―――― *On the Face on the Moon*, trans. Harold Cherniss, in *Plutarch's Moralia* 12, trans. Harold Cherniss and William C. Helmbold, Cambridge, MA: Harvard University Press (*LCL*), 1957/1984.

―――― *On the Festival of Images at Plataea*, in *Plutarch's Moralia* 15 *Fragments*, trans. F. H. Sandbach, Cambridge, MA: Harvard University Press (*LCL*), 1969, 285-87.

―――― *On Listening to Lectures* (*De audiendo*) *Plutarch's Moralia* 1, trans. Frank Cole Babbitt, Cambridge, MA: Harvard University Press (*LCL*), 1931.

―――― *Table Talk* in *Plutarch's Moralia* 8 & 9, trans. P. A. Clement and H. B. Hoffleit; E. L. Minar, F. H. Sandbach and W. C. Helmbold, Cambridge, MA: Harvard University Press (*LCL*), 1969, 1961.

Porphyry *On the Life of Plotinus* in *Plotinus* I: *Porphyry On the Life of Plotinus and the Order of his Books; Enneads I. 1-9*, trans. A. H. Armstrong, London: William Heinemann (*LCL*), 1966.

[Ptolemy] *Ptolemy's* Almagest, trans. G. J. Toomer, New York: Springer-Verlag, 1984.

Scriptores Historiae Augustae, 3 vols, trans. David Magie, London: Heinemann (*LCL*), 1921-32, vol. 1.

Seneca *Letters to Lucilius (Ad Lucilium epistulae morales)*, trans. R. M. Gummere, in *Seneca, Epistles*, 3 vols, Cambridge, MA: Harvard University Press (*LCL*), 1917-1925.

―――― *Natural Questions*, trans. T. H. Corcoran, *Seneca, Naturales Quaestiones*, 2 vols, London: William Heinemann (*LCL*), 1971-72.

―――― *On Benefits* (*De beneficiis*) in *Seneca in Ten Volumes, Moral Essays*, vol. 3, trans. John W. Basore, Cambridge, MA: Harvard University Press, (*LCL*), 1935.

[Simplicius] *Simplicii in Aristotelis Physicorum libros quattuor priores commentaria*, ed. Hermann Diels, Berlin: Reimer, 1882, *Commentaria in Aristotelem graeca* 9 (= *Commentary on Aristotle's Physics*).

―――― *On Aristotle's* Physics 2, trans. Barrie Fleet. Ithaca, NY: Cornell University Press, 1997.

Stoicorum Veterum Fragmenta, ed. H. von Arnim. 4 vols. Leipzig: B.G. Teubner, 1903-24.

Strabo *The Geography of Strabo*, 8 vols, trans. H. L. Jones, Cambridge, MA: Harvard University Press (*LCL*), 1917-32.

Theophrastus *Meteorology*, in H. Daiber (1992) "The Meteorology of Theophrastus in Syriac and Arabic Translation," in Fortenbaugh,

W. W., and D. Gutas, (eds.) *Theophrastus: His Psychological, Doxographical, and Scientific Writings*, New Brunswick, NJ: Transaction Publishers, *Rutgers University Studies in Classical Humanities*, vol. 5, 166-293.

Modern authors

Asmis, E. (1984) *Epicurus' Scientific Method.* Ithaca, NY: Cornell University Press.

Ausland, Hayden (1997) "On Reading Plato Mimetically," *American Journal of Philology* 118: 371-416.

Bagnall, Roger S. (1995) *Reading Papyri, Writing Ancient History.* London: Routledge.

Barnes, Jonathan (1995) "Life and Work," in Barnes, Jonathan (ed.), *The Cambridge Companion to Aristotle.* Cambridge, England: Cambridge University Press, 1-26.

——— (1997) "Roman Aristotle," in Barnes, Jonathan, and Miriam Griffin (eds.), *Philosophia Togata II.* Oxford, England: Clarendon Press, 1-69.

Beazley, J. D. (1984) "Hymn to Hermes," *American Journal of Archaeology* 52 (1948):336-40.

Betegh, Gábor (2004) *The Derveni Papyrus: Cosmology, Theology and Interpretation.* Cambridge, England: Cambridge University Press.

Blair, Ann (2000) "Annotating and Indexing Natural Philosophy," in Frasca-Spada, Marina, and Nick Jardine (eds.), *Books and the Sciences in History.* Cambridge, England: Cambridge University Press, 79-89.

Blondell, Ruby (2002) *The Play of Character in Plato's Dialogues.* Cambridge, England: Cambridge University Press.

Boys-Stones, G. R. (2001) *Post-Hellenistic Philosophy: A Study of its Development from the Stoics to Origen.* Oxford, England: Oxford University Press.

Broadie, Sarah (2001) "Theodicy and Pseudo-history in the Timaeus," *Oxford Studies in Ancient Philosophy* 21: 1-28.

Buitron-Oliver, Diana et al. (1992) *The Greek Miracle: Classical Sculpture from the Dawn of Democracy, The Fifth Century B.C.* Washington, DC: National Gallery of Art.

Burkert, Walter (1972) *Lore and Science of Ancient Pythagoreanism,* trans. Edwin L. Minar, Jr., Cambridge, MA: Harvard University

Press, original publication *Weisheit und Wissenschaft: Studien zu Pythagoras, Philolaos und Platon*. Nürnberg: Verlag Hans Carl, 1962.

Burns, Alfred (1981) "Athenian Literacy in the Fifth Century B.C.," *Journal of the History of Ideas* 42: 371-87.

Burnyeat, M. F. (2000) "Why Mathematics is Good for the Soul," in Timothy Smiley (ed.), *Mathematics and Necessity: Essays in the History of Philosophy, Proceedings of the British Academy* 103: 1-81.

Buxton, Richard (ed.) (1999) *From Myth to Reason?: Studies in the Development of Greek Thought*. Oxford, England: Oxford University Press.

Calame, Claude (1999) "The Rhetoric of *Muthos* and *Logos*: Forms of Figurative Discourse," in Buxton, Richard (ed.), *From Myth to Reason?: Studies in the Development of Greek Thought*. Oxford, England: Oxford University Press, 119-43.

Cameron, Alan (1967) "The End of the Ancient Universities," *Cahiers d'Histoire Mondiale* 10: 653-73.

Campbell, T. M. (1908) "Longfellow and the Hexameter," *Modern Language Notes* 23, 3: 96.

Cavallo, Guglielmo (1999) "Between *Volumen* and Codex," in Cavallo, Guglielmo, and Roger Chartier (eds.), *A History of Reading in the West*, trans. Lydia G. Cochrane, Cambridge, England: Polity Press, 64-89.

———, and Chartier, Roger (eds.) (1999) *A History of Reading in the West*, trans. Lydia G. Cochrane, Cambridge, England: Polity Press.

Cherniss, Harold (1957/1984) "Introduction" to Plutarch *On the Face on the Moon*, in *Plutarch's Moralia* 12, trans. Harold Cherniss and William C. Helmbold, Cambridge, MA: Harvard University Press (*LCL*), 1957/1984, 2-33.

———. (1976/2000) "Introduction" to Plutarch *Platonic Questions*, in *Plutarch's Moralia*. Cambridge, MA: Harvard University Press (*LCL*), vol. 13, part 1.

Clay, Diskin (1983) *Lucretius and Epicurus*. Ithaca, NY: Cornell University Press.

Conte, Gian Biagio (1994) *Genres and Readers: Lucretius, Love Elegy, Pliny's Encyclopedia*. trans. Glenn W. Most. Baltimore, MD: The Johns Hopkins University Press.

———, and Glenn W. Most (1996) "Genre," in Hornblower, Simon, and Antony Spawforth (eds.), *The Oxford Classical Dictionary*, 3rd edn. Oxford, England: Oxford University Press.

Coones, P. (1983) "The Geographical Significance of Plutarch's Dialogue, concerning the Face Which Appears in the Orb of the Moon," *Transactions of the Institute of British Geographers* 8 (1983): 361-72.

Cornford, F. M. (1912/1957) *From Religion to Philosophy: A Study in the Origins of Western Speculation*. New York: Harper & Row.

——— (1937/1975) *Plato's Cosmology: The* Timaeus *of Plato,* translated with a running commentary. Indianapolis, IN: Bobbs-Merrill; originally published London: Routledge and Kegan Paul.

Cuomo, Serafina (2001) *Ancient Mathematics*. London: Routledge.

——— (2007) *Technology and Culture in Greek and Roman Antiquity*. Cambridge, England: Cambridge University Press.

Darwin, Erasmus (1789) *The Botanic Garden: a Poem, in two parts ... with philosophical notes*. London: For J. Johnson (pt. 2: J. Jackson, Lichfield, sold by J. Johnson), 1789-91.

De Leemans, P., and M. Goyens (eds.) (2006) *Aristotle's* Problemata *in Different Times and Tongues*. Leuven, Netherlands: Leuven University Press.

Depew, Mary, and Dirk Obbink (eds.) (2000) *Matrices of Genre: Authors, Canons, and Society*. Cambridge, MA: Harvard University Press.

Dick, Steven J. (1980) "The Origins of the Extraterrestrial Life Debate and its relation to the Scientific Revolution," *Journal of the History of Ideas* 41: 3-27.

——— (1982) *Plurality of Worlds: The Origins of the Extraterrestrial Life Debate from Democritus to Kant*. Cambridge, England: Cambridge University Press.

Dirlmeier, Franz (1962) *Merkwürdige Zitate in der Eudemischen Ethik des Aristoteles*. Heidelberg, Germany: Carl Winter, Universitätsverlag.

Duff, David (ed.) (2000) *Modern Genre Theory*. Harlow, England: Longman.

Edelstein, L. (1967) *The Idea of Progress in Classical Antiquity*. Baltimore, MD: The Johns Hopkins University Press.

Ferrari, G. R F. (1984) "Orality and Literacy in the Origin of Philosophy," review of Kevin Robb (ed.), *Language and Thought in Early Greek Philosophy*. Lasalle, IL: Hegeler Institute, 1983, in *Ancient Philosophy* 4:194-205.

Flashar, Helmut (trans.) (1975) *Aristoteles: Problemata physica*. Berlin: Akademie Verlag.

Fögens, Thorsten (2005) *Antike Fachtexte: Ancient Technical Texts*. Berlin: Walter de Gruyter.

Fowler, P. G., and D. P. Fowler (1997/1999) "Introduction" to *Lucretius: On the Nature of the Universe*, trans. Ronald Melville. Oxford, England: Oxford University Press, Oxford's World Classics; first published 1997 by Clarendon Press.

Frasca-Spada, Marina, and Nick Jardine (eds.) (2000) *Books and the Sciences in History*. Cambridge, England: Cambridge University Press.

Furley, David (1970/1989) "Variations on Themes from Empedocles in Lucretius' Proem," *Bulletin of the Institute of Classical Studies* 17: 55-64, reprinted in Furley, D. J. (1989) *Cosmic Problems*. Cambridge, England: Cambridge University Press, 172-82.

——— (1999) "Cosmology" in Algra, K., J. Barnes, J. Mansfeld, and M. Schofield (eds.), *The Cambridge History of Hellenistic Philosophy*. Cambridge, England: Cambridge University Press, 412-51.

Fuhrmann, Manfred (1960) *Das systematische Lehrbuch: Ein Beitrag zur Geschichte der Wissenschaften in der Antike*. Göttingen, Germany: Vandenhoeck & Ruprecht.

Gadamer, Hans-Georg (2004) *Truth and Method*, 2nd revised edn, translation revised by Joel Weinsheimer and Donald G. Marshall. London/New York: Continuum, originally published as *Wahrheit und Methode: Grundzüge einer philosophischen Hermeneutik*. Tübingen, Germany: Mohr, 1960.

Galilei, Galileo (1967) *Dialogue Concerning the Two Chief World Systems*, trans. Stillman Drake. Berkeley: University of California Press, 2nd edn.

Gernet, Louis (1983) *Les Grecs sans Miracle*, ed. R. di Donato. Paris: La Découverte.

Gibson, Roy K., and Christina Shuttleworth Kraus (eds.) (2002) *The Classical Commentary: Histories, Practices, Theory*. Leiden, Netherlands: Brill.

Goodyear, F. R. D. (ed.) (1965) *Incerti Auctoris Aetna*, with an introduction and commentary. Cambridge, England: Cambridge University Press.

——— (1984) "The 'Aetna': Thought, Antecedents, and Style," in *Aufstieg und Niedergang der römischen Welt (ANRW)*. Berlin: de Gruyter, 2.32.1: 344-63.

Goold, G. P. (1977) *Manilius: Astronomica*. Cambridge, MA: Harvard University Press (*LCL*).

Graf, Fritz (1996) "Pythagoras, Pythagoreanism," in Hornblower, Simon, and Antony Spawforth (eds.) (1996) *The Oxford Classical Dictionary*, 3rd edn. Oxford, England: Oxford University Press.

Grafton, Anthony (1997) *Commerce with the Classics: Ancient Books and Renaissance Readers*. Ann Arbor, MI: University of Michigan Press.

——— (2000) "Geniture Collections, Origins and Uses of a Genre," in Frasca-Spada, Marina, and Nick Jardine (eds.), *Books and the Sciences in History*. Cambridge, England: Cambridge University Press, 49-68.

———, and Megan Williams (2006) *Christianity and the Transformation of the Book*. Cambridge, MA: Harvard University Press.

Grant, Michael, and Rachel Kitzinger (eds.) (1981) *Civilization of the Ancient Mediterranean: Greece and Rome*. New York: Charles Scribner's Sons, vol. 1.

Guthrie, W. K. C. (1962) *A History of Greek Philosophy: The Earlier Presocratics and the Pythagoreans*. Cambridge, England: Cambridge University Press, vol. 1.

Hahm, D. E. (1977) *The Origins of Stoic Cosmology*. Columbus, OH: Ohio State University Press.

Harris, William V. (1989) *Ancient Literacy*. Cambridge, MA: Harvard University Press.

Harvey, Paul (1937/1974) *Oxford Companion to Classical Literature*. Oxford, England: Clarendon Press; first published in 1937, and corrected in subsequent printings.

Haslam, Michael (1972) "Plato, Sophron and the Dramatic Dialogue," *Bulletin of the Institute of Classical Studies* 19: 17-38.

Havelock, Eric A. (1963) *Preface to Plato*. Oxford, England: Blackwell.

——— (1966) "Pre-literacy and the Pre-Socratics," *Bulletin of the Institute of Classical Studies* 13: 44-67.

——— (1971) *Prologue to Greek Literacy*. Cincinnati, OH: University of Cincinnati Press.

——— (1976) *Origins of Western Literacy: Four Lectures Delivered at the Ontario Institute for Studies in Education, Toronto, March 25, 26, 27, 28, 1974*. Toronto: Ontario Institute for Studies in Education Monograph Series 14.

——— (1982) *The Literate Revolution in Greece and its Cultural Consequences*. Princeton, NJ: Princeton University Press.

——— (1986) *The Muse Learns to Write: Reflections on Orality and Literacy from Antiquity to the Present*. New Haven, CT: Yale University Press.

Hett, W. S. (1957) "Introduction," Aristotle *Problems*, 2 vols. London: Heinemann (*LCL*), vol. 1.

Hine, H. M. (1981) *An Edition with Commentary of Seneca.* Natural Questions, *Book Two*, New York: Arno Press.

——— (1996) "*Aetna*," in Hornblower, Simon, and Antony Spawforth (eds.), *The Oxford Classical Dictionary*, 3rd edn. Oxford, England: Oxford University Press.

——— (2002) "Seismology and Vulcanology in Antiquity," in Tuplin, C. J. and T. E. Rihll (eds.) *Science and Mathematics in Ancient Greek Culture.* Oxford, England: Oxford University Press, 56-75.

Hirsch, E. (1967) *Validity in Interpretation.* New Haven, CT: Yale University Press.

Hornblower, Simon, and Antony Spawforth (eds.) (1996) *The Oxford Classical Dictionary*, 3rd edn. Oxford, England: Oxford University Press

Hunter, Andrew (ed.) (2000) *Scientific Books, Libraries and Collectors*, 4th edn. Aldershot, England: Ashgate.

Hunter, Richard (1996) *Theocritus and the Archaeology of Greek Poetry.* Cambridge, England: Cambridge University Press.

Immerwahr, Henry (1964) "Book Rolls on Attic Vases," in Charles Henderson, Jr. (ed.) *Classical, Mediaeval and Renaissance Studies in Honor of Berthold Louis Ullman*, 2 vols in 1. Rome: Edizioni di Storia e Letteratura, 1. 17-48.

Inwood, M. J. (1992) "Problematic Problems," review of Pierre Louis (ed. and trans.) (1991) *Aristote*, Problèmes, I, *Sections I à X*. Paris: Les Belles Lettres, in *Classical Review* 42, 2: 285-86.

Jardine, Nicholas (1991) "Writing off the Scientific Revolution," *Journal for the History of Astronomy* 22: 311-18.

———, and Alain Segonds (1999), "Kepler as Reader and Translator of Aristotle," in Blackwell, Constance, and Sachiko Kusukawa (eds.), *Philosophy in the Sixteenth and Seventeenth Centuries: Conversations with Aristotle.* Aldershot, England: Ashgate, 206-33.

Johansen, Thomas K. (1999) "Myth and *Logos* in Aristotle," in *From Myth to Reason?: Studies in the Development of Greek Thought.* Richard Buxton (ed.), Oxford, England: Oxford University Press, 279-91.

——— (2004) *Plato's Natural Philosophy: A Study of the Timaeus-Critias*.Cambridge, England: Cambridge University Press.

Johnson, William A. (2000) "Toward a Sociology of Reading in Classical Antiquity," *American Journal of Philology* 121: 593-627.

Kahn, Charles H. (1996) *Plato and the Socratic Dialogue: The Philosophical Use of a Literary Form*. Cambridge, England: Cambridge University Press.

Kenney, E. J. (1970) "Doctus Lucretius," *Mnemosyne* 23: 366-92.

Kepler, Johannes (1870) *Opera Omnia*, Christian Frisch (ed.). Frankfurt am Main, Germany: Heyder & Zimmer, vol. 8.

——— (1945), *Briefe 1590-1599* in Caspar, Max (ed.). *Gesammelte Werke*, Bd. XIII, Munich, Germany: C.H. Beck, 1938-88.

[Kepler, Johannes] (1951), *Johannes Kepler: Life and Letters*, trans. Carola Baumgardt. New York: Philosophical Library.

——— (1967) *Somnium: the Dream, or Posthumous Work on Lunar Astronomy*, trans. Edward Rosen. Madison: University of Wisconsin Press.

Kidd, Douglas (1997) *Aratus: Phaenomena*. Cambridge, England: Cambridge University Press.

Kirk, G. S., J. E. Raven, and M. Schofield (1983) *The Presocratic Philosophers*, 2nd edn. Cambridge, England: Cambridge University Press.

König, Jason, and Tim Whitmarsh (eds.) (2007) *Ordering Knowledge in the Roman Empire*. Cambridge, England: Cambridge University Press.

Konstan, David (1991) "What is Greek about Greek Mythology?," *Kernos* 4: 11-30.

——— (2004) " 'The Birth of the Reader': Plutarch as Literary Critic," *Scholia* 13 (2004) 3-27.

——— (2005) "Introduction," in Russell, Donald A., and David Konstan (eds. and trans.), *Heraclitus:* Homeric Problems. Leiden, Netherlands: Brill, xi-xxx.

——— (2007a) "Cartesian Solitude and the Nature of Reading." Unpublished paper, presented at the conference, "Philosophy and Literature: Reading across the Disciplines," held at Wesleyan University, CT, on May 9-10, 2007.

——— (2007b) "The Active Reader and the Ancient Novel." Unpublished paper, presented at the conference, "Readers and Writers in the Ancient Novel," 4th Rethymnon International Conference on the Ancient Novel, held at the University of Crete at Rethymno on May 20-22, 2007.

Lamberton, Robert (1992) "Introduction," in Lamberton, Robert, and John J. Keaney, *Homer's Ancient Readers: The Hermeneutics of Greek*

Epic's Earliest Exegetes. Princeton, NJ: Princeton University Press, vii-xxiv.

——— (2001) *Plutarch*. New Haven, CT: Yale University Press.

———, and John J. Keaney (1992) *Homer's Ancient Readers: The Hermeneutics of Greek Epic's Earliest Exegetes*. Princeton, NJ: Princeton University Press.

Lewenstein, Bruce V. (1992) "Cold Fusion and Hot History," *Osiris* (2nd series) 7: 135-63.

Lewis, Naphtali (1974) *Papyrus in Classical Antiquity*. Oxford, England: Clarendon Press, an enlarged edition of *L'Industrie du Papyrus dans l'Égypte Gréco-romaine*. Paris, 1934.

Liddell, H. G., R. Scott, and H. S. Jones (1940/1968) *A Greek-English Lexicon*, 9th edn. Oxford, England: Clarendon Press, with *Supplement* by Barber, E. A. *et al.*, reprinted in 1 vol.

Lindberg, David C. (1992) *The Beginnings of Western Science: The European Scientific Tradition in Philosophical, Religious and Institutional Context, 600 B.C. to A.D. 1450*. Chicago: University of Chicago Press.

Lloyd, G. E. R. (1966) *Polarity and Analogy: Two Types of Argumentation in Early Greek Thought*. Cambridge, England: Cambridge University Press.

——— (1967/1991) "Popper versus Kirk: A Controversy in the Interpretation of Greek Science," reprinted in Lloyd (1991), 100-120.

——— (1975/1991) "Greek Cosmologies," reprinted in Lloyd (1991), 141-63.

——— (1987) *The Revolutions of Wisdom: Studies in the Claims and Practice of Ancient Greek Science*. Berkeley: University of California Press.

——— (1991) *Methods and Problems in Greek Science: Selected Papers*. Cambridge, England: Cambridge University Press.

Lloyd, Geoffrey, and Nathan Sivin (2002) *The Way and the Word: Science and Medicine in Early China and Greece*. New Haven, CT: Yale University Press.

Lloyd-Morgan, Glenys (1996) "Mirrors," in Hornblower, Simon, and Antony Spawforth (eds.), *The Oxford Classical Dictionary*, 3rd edn. Oxford, England: Oxford University Press.

Long, A. A. (1992) "Stoic Readings of Homer," in Lamberton, Robert, and John J. Keaney, *Homer's Ancient Readers: The Hermeneutics of Greek Epic's Earliest Exegetes*. Princeton, NJ: Princeton University Press, 41-66.

———, and D. N. Sedley (1987) *The Hellenistic Philosophers*, 2 vols. Cambridge, England: Cambridge University Press.
Longfellow, Henry W. (1841/1848) *Evangeline: A Tale of Acadie*. London: H. G. Clarke and Co.
Lord, Albert B. (1960) *The Singer of Tales*. Cambridge, MA: Harvard University Press.
Louis, Pierre (ed. and trans.) (1991) *Aristote*, Problèmes, I, *Sections I à X*. Paris: Les Belles Lettres, 3 vols.
Lowenstam, Steven (1997) "Talking Vases: The Relationship between the Homeric Poems and Archaic Representations of Epic Myth," *Transactions of the American Philological Association* 127: 21-76.
Mansfeld, J. (1986/1990) "Aristotle, Plato, and the Pre-platonic Doxography and Chronography," in Mansfeld, J., *Studies in the Historiography of Greek Philosophy*. Assen/Maastricht, Netherlands: Van Gorcum, 22-83, reprinted from G. Cambiano (ed.) (1986) *Storiografia e dossografia nella filosofia antica*. Turin, Italy: Tirrenia Stampatori, 1-59.
——— (1992) *"Physikai Doxai* and *Problēmata physika* from Aristotle to Aëtius (and Beyond)," in Fortenbaugh, W. W., and D. Gutas, *Theophrastus: His Psychological, Doxographical and Scientific Writings*. New Brunswick, NJ: Transaction, Rutgers University Studies in Classical Humanities, vol. 5, 63-111.
——— (1994) *Prolegomena: Questions to be Settled before the Study of an Author, or a Text*. Leiden, Netherlands: E. J. Brill.
——— (1999) "Theology," in Algra, K., J. Barnes, J. Mansfeld, and M. Schofield (eds.) *The Cambridge History of Hellenistic Philosophy*. Cambridge, England: Cambridge University Press, 452-78.
———, and D. Runia (1997) *Aëtiana: The Method and Intellectual Context of a Doxographer*, vol. 1. *The Sources*, Leiden, Netherlands: Brill.
McLuhan, Marshall (1964) *Understanding Media: The Extensions of Man*. New York: McGraw-Hill.
Morgan, Kathryn A. (2000) *Myth and Philosophy from the Presocratics to Plato*. Cambridge, England: Cambridge University Press.
Most, Glenn W. (1999a) "From Logos to Mythos," in Buxton, Richard (ed.), *From Myth to Reason?: Studies in the Development of Greek Thought*. Oxford, England: Oxford University Press, 25-47.
——— (1999b) "The Poetics of Early Greek Philosophy," in *The Cambridge Companion to Early Greek Philosophy*, A. A. Long (ed.). Cambridge, England: Cambridge University Press, 332-62.

———— (ed.) (1999c) *Commentaries—Kommentare*. Göttingen, Germany: Vandenhoeck & Ruprecht.

————, and Gian Biagio Conte (1996) "*Topos*," in Hornblower, Simon, and Antony Spawforth (eds.). *The Oxford Classical Dictionary*, 3rd edn, Oxford, England: Oxford University Press.

Mourelatos, Alexander P. D. (1970) *The Route of Parmenides: A Study of Word, Image and Argument in the Fragments*. New Haven, CT, and London: Yale University Press.

Murray, Penelope (1999) "What is a *Muthos* for Plato?," in Buxton, Richard (ed.), *From Myth to Reason?: Studies in the Development of Greek Thought*. Oxford, England: Oxford University Press, 251-62.

Nestle, Wilhelm (1942) *Von Mythos zum Logos: Die Selbstentfaltung des griechischen Denkens von Homer bis auf die Sophistik und Sokrates*, 2nd edn. Stuttgart, Germany: Kröner; 1st edn published 1940.

Netz, Reviel (1999) *The Shaping of Deduction in Greek Mathematics: A Study in Cognitive History*. Cambridge, England: Cambridge University Press.

Nightingale, Andrea Wilson (1996) *Genres in Dialogue: Plato and the Construct of Philosophy*. Cambridge, England: Cambridge University Press.

Nussbaum, M. (1996) "Aristotle," in Hornblower, Simon, and Antony Spawforth (eds.), *The Oxford Classical Dictionary*, 3rd edn. Oxford, England: Oxford University Press.

Obbink, Dirk (ed.) (1995) *Philodemus and Poetry: Poetic Theory and Practice in Lucretius, Philodemus and Horace*. New York: Oxford University Press.

———— (1996) "Introduction," to *Philodemus* On Piety: *Part I*. Oxford, England: Clarendon Press.

Osborne, C. (1996) "Space, Time, Shape and Direction: Creative Discourse in the *Timaeus*," in *Form and Argument in Late Plato*, Gill, C., and M. M. McCabe (eds.). Cambridge, England: Cambridge University Press, 179-211.

Paisley, P. B., and D. R. Oldroyd (1979) "Science in the Silver Age: *Aetna*, a Classical Theory of Volcanic Activity," *Centaurus* 23: 1-20.

Parker, Robert (1996) "Hymns (Greek)," in Hornblower, Simon, and Antony Spawforth (eds.), *The Oxford Classical Dictionary*, 3rd edn. Oxford, England: Oxford University Press.

Parry, Millman (1971) *The Making of Homeric Verse: the Collected Papers of Milman Parry*, Parry, Adam (ed.). Oxford, England: Clarendon Press.

Paton, W. R. (trans.) *Greek Anthology*, 5 vols. Cambridge, MA: Harvard University Press, (*LCL*), 1916-19.

Peters, F. E. (1967) *Greek Philosophical Terms: A Historical Lexicon*. New York: New York University Press.

Pfeiffer, Rudolph (1968/1978) *History of Classical Scholarship from the Beginning to the End of the Hellenistic Age*. Oxford, England: Clarendon Press.

Platthy, Jeno (1968) *Sources on the Earliest Greek Libraries with the Testimonia*. Amsterdam, Netherlands: Hakkert.

Ramelli, Ilaria, and Giulio Lucchetta (2004) *Allegoria*. Vol. 1: *L'età classica*. Milan, Italy: V&P Università.

Reardon, B. P. (ed.) (1989) *Collected Ancient Greek Novels*. Berkeley: University of California Press.

Reymond, Arnold (1927) *History of the Sciences in Greco-Roman Antiquity*. trans. Ruth Gheury de Bray, London: Methuen.

Reynolds, L. D. and N. G. Wilson (1991) *Scribes and Scholars: A Guide to the Transmission of Greek and Latin Literature*, 3rd ed. Oxford, England: Clarendon Press.

Richardson, N. J. (1992) "Aristotle's Reading of Homer and its Background," in Lamberton, Robert, and John J. Keaney (eds.) *Homer's Ancient Readers: The Hermeneutics of Greek Epic's Earliest Exegetes*. Princeton, NJ: Princeton University Press, 30-40.

Roberts, Colin H., and T. C. Skeat (1983) *The Birth of the Codex*. London: Oxford University Press for the British Academy.

Ross, W. D. (1924) *Aristotle's* Metaphysics. Oxford, England: Clarendon Press, 2 vols.

Rowe, Christopher (1996) "Dialogue, Greek" in Hornblower, Simon, and Antony Spawforth (eds.), *The Oxford Classical Dictionary*, 3rd edn. Oxford, England: Oxford University Press. [While the initials used to sign the article—C. Ro.—do not correspond to any of the abbreviations listed for authors, authorship was confirmed with Rowe, through private correspondence.]

——— (1999) "Myth, History, and Dialectic in Plato's *Republic* and *Timaeus-Critias*," in Buxton, Richard (ed.) *From Myth to Reason?: Studies in the Development of Greek Thought*. Oxford, England: Oxford University Press, 263-78.

Runia, David T. (1999) "What is Doxography?," in van der Eijk, Philip J. (ed.) *Ancient Histories of Medicine: Essays in Medical Doxography and Historiography in Classical Antiquity*. Leiden, Netherlands: E. J. Brill, *Studies in Ancient Medicine* 20, 33-55.

Rupprecht, H.-A. (1994) *Kleine Einführung in die Papyruskunde*. Darmstadt, Germany: Wissenschaftliche Buchgesellschaft.

Russell, Donald A., and David Konstan (eds. and trans.) (2005) *Heraclitus:* Homeric Problems. Leiden, Netherlands: Brill.

Sambursky, S. (1959) *Physics of the Stoics*. Princeton, NJ: Princeton University Press.

Sarton, George (1952/1993) *Ancient Science through the Golden Age of Greece*. New York: Dover; London: Constable; originally published as vol. 1 of *A History of Science*. Cambridge, MA: Harvard University Press.

Scarborough, John (1982) "Beans, Pythagoras, Taboos and Ancient Dietetics," *Classical World* 75:355-58.

Schiesaro, Alessandro (1996) "Didactic Poetry," in Hornblower, Simon, and Antony Spawforth (eds.), *The Oxford Classical Dictionary*, 3rd edn. Oxford, England: Oxford University Press.

Schiesaro, A., P. Mitsis, and J. S. Clay (eds.) (1993) *Mega nepios: il destinatario nell'epos didascalico*. Pisa, Italy: Giardini, *Materiali e discussioni per l'analisi dei testi classici* 31.

Schrader, Hermann (1880) *Porphyrii Quaestionum homericarum ad Iliadem*. Leipzig, Germany: B.G. Teubner.

Sedley, David (1976) "Epicurus and the Mathematicians of Cyzicus," *Cronache Ercolanesi* 6: 23-54.

——— (1990) "Teleology and Myth in the *Phaedo*," *Proceedings of the Boston Area Colloquium in Ancient Philosophy* 5: 359-83.

——— (1998) *Lucretius and the Transformation of Greek Wisdom*. Cambridge, England: Cambridge University Press.

——— (2007) *Creationism and its Critics in Antiquity*. Berkeley: University of California Press.

Sharples, R. W. (2006) "*Pseudo-Alexander or pseudo-Aristotle, Medical Puzzles and Physical Problems*," in De Leemans, P., and M. Goyens (eds), *Aristotle's* Problemata *in Different Times and Tongues*. Leuven, Netherlands: Leuven University Press, 21-31.

Sider, David (2005) *The Fragments of Anaxagoras*. 2nd edn. Sankt Augustin, Germany: Academia Verlag.

Sluiter, Ineke (2000) "The Dialectics of Genre: Some Aspects of Secondary Literature and Genre in Antiquity," in Depew, Mary, and Dirk Obbink (eds.), *Matrices of Genre: Authors, Canons, and Society*. Cambridge, MA: Harvard University Press, 183-203.

Small, Jocelyn Penny (1997) *Wax Tablets of the Mind: Cognitive Studies of Memory and Literacy in Classical Antiquity*. London: Routledge.

Smith, Robin (1997) *Topics: Books I and VIII*, with excerpts from related texts: translated with a commentary. Oxford, England: Clarendon Press.

Snell, Bruno (1953) *The Discovery of the Mind: The Greek Origins of European Thought*, trans. T. G. Rosenmeyer. Oxford, England: Basil Blackwell; originally published as *Die Entdeckung des Geistes: Studien zur Entstehung des europäischen Denkens bei den Griechen*. Hamburg, Germany: Claaszen und Govert, 1946.

Snyder, Gregory (2000) *Teachers and Texts in the Ancient World: Philosophers, Jews and Christians*. London: Routledge.

Solmsen, Friedrich (1953) "Epicurus on the Growth and Decline of the Cosmos," *American Journal of Philology* 74: 34-51.

——— (1957) "Lucretius 1.174 and Aetna 1," *Classical Philology* 52: 251.

Stephens, Susan A. (1981) "Book Production," in Grant, Michael, and Rachel Kitzinger (eds.), *Civilization of the Ancient Mediterranean: Greece and Rome*. New York: Charles Scribner's Sons, vol. 1.

Sudhaus, Siegfried (1898) *Aetna*. Leipzig, Germany: B. G. Teubner.

Taub, Liba (1993) *Ptolemy's Universe: The Natural Philosophical and Ethical Foundations of Ptolemy's Astronomy*. Chicago, IL: Open Court.

——— (2000) "Ancient Science," in Hunter, Andrew (ed.) *Scientific Books, Libraries and Collectors*, 4th edn. Aldershot, England: Ashgate, pp. 26-71.

——— (2003) *Ancient Meteorology*, London: Routledge.

——— (2007) "Presenting a 'Life' as a Guide to Living: Ancient Accounts of the Life of Pythagoras," in Söderqvist, Thomas (ed.), *The History and Poetics of Scientific Biography*. Aldershot, England: Ashgate, 17-36.

——— (2008) "'Eratosthenes Sends Greetings to King Ptolemy': Reading the Contents of a 'Mathematical' Letter," in Dauben, Joseph W., Stefan Kirschner, Andreas Kühne, Paul Kunitzsch, and Richard P. Lorch (eds.), *Mathematics Celestial and Terrestrial: Festschrift für Menso Folkerts zum 65. Geburtstag*. Halle/Saale: Deutsche Akademie der Naturforscher Leopoldina, *Acta Historica Leopoldina*, 54.

——— (forthcoming) "Epicurean Cosmology and Meteorology," in Warren, James (ed.), *The Cambridge Companion to Epicureanism*. Cambridge, England: Cambridge University Press.

Taylor, A. E. (1928) *A Commentary on Plato's Timaeus*. Oxford, England: Clarendon Press.

Thalmann, William G. (1984) *Conventions of Form and Thought in Early Greek Epic Poetry.* Baltimore, MD: The Johns Hopkins University Press.

Todorov, Tzvetan (1974) "Literary Genres," in Sebeok, Thomas A. (ed.) *Current Trends in Linguistics,* 14 vols, vol. 12 part 2, *Linguistics and Adjacent Arts and Sciences.* The Hague, Netherlands: Mouton, 957-62.

Toohey, P. (1996) *Epic Lessons: An Introduction to Ancient Didactic Poetry.* London: Routledge.

Toomer, G. J. (1996a) "Aratus," in Hornblower, Simon, and Antony Spawforth (eds.), *The Oxford Classical Dictionary,* 3rd edn. Oxford, England: Oxford University Press.

—— (1996b) "Nicomachus (3)," in Hornblower, Simon, and Antony Spawforth (eds.), *The Oxford Classical Dictionary,* 3rd ed. Oxford, England: Oxford University Press.

Trapp, Michael B. (1996) "Letters, Greek," in Hornblower, Simon, and Antony Spawforth (eds.), *The Oxford Classical Dictionary,* 3rd edn. Oxford, England: Oxford University Press.

Turner, E. G. (1968/1980) *Greek Papyri.* Oxford, England: Clarendon Press.

Turner, Eric G. (1977) *The Typology of the Early Codex.* [Philadelphia]: University of Pennsylvania Press.

Van der Eijk, Philip J. (1997) "Towards a Rhetoric of Ancient Scientific Discourse: Some Formal Characteristics of Greek Medical and Philosophical Texts (Hippocratic Corpus, Aristotle)," in Bakker, Egbert J. (ed.), *Grammar as Interpretation: Greek Literature in its Linguistic Contexts.* Leiden, Netherlands: Brill, 77–129.

Vernant, Jean-Pierre (1982) *The Origins of Greek Thought.* Ithaca, NY: Cornell University Press. Originally published as *Les origines de la pensée grecque.* Paris: Presses universitaires de France, 1962; 2nd edn 1992.

—— (1983) *Myth and Thought among the Greeks.* London: Routledge & Kegan Paul; originally published as *Mythe et pensée chez les Grecs.* Paris: F. Maspero, 1965.

—— (1995) *Passé et présent: Contributions à une psychologie historique réunies par Riccardo di Donato.* Rome: Edizioni di storia e letteratura.

Volk, Katherina (2002) *The Poetics of Latin Didactic: Lucretius, Vergil, Ovid, Manilius.* Oxford, England: Oxford University Press.

—— (2005) "*Aetna* oder wie man ein Lehrgedicht schreibt," in Holzberg, N. (ed.), *Die Appendix Vergiliana: Pseudepigraphen im literarischen Kontext*. Tubingen, Germany: Gunter Narr Verlag, 68-89.

Von Fritz, Kurt, and Ernst Kapp (1950/1974) *Aristotle's* Constitution of Athens *and Related Texts*. New York: Hafner Press.

Von Staden, Heinrich (1992) "Affinities and Elisions: Helen and Hellenocentrism," *Isis* 83: 578-95.

——— (2002) "'A Woman does not become ambidextrous': Galen and the Culture of Scientific Commentary," in Gibson, Roy K., and Christina Shuttleworth Kraus (eds.), *The Classical Commentary: Histories, Practices, Theories*. Leiden: Brill, 109-38.

Warren, James (2007) "Diogenes Laërtius, Biographer of Philosophy," in König, Jason, and Tim Whitmarsh (eds), *Ordering Knowledge in the Roman Empire*. Cambridge, England: Cambridge University Press, 133-49.

West, M. L. (1996) "Hesiod," in Hornblower, Simon, and Antony Spawforth (eds.), *The Oxford Classical Dictionary*, 3rd edn. Oxford, England: Oxford University Press.

Williams, G. D. (2006) "Greco-Roman Seismology and Seneca on Earthquakes in Natural Questions 6," *Journal of Roman Studies* 96: 124-46.

Woodhead, A. G. and Roger J. A. Wilson (1996) "Aetna (1)," in Hornblower, Simon, and Antony Spawforth (eds.), *The Oxford Classical Dictionary*, 3rd edn. Oxford, England: Oxford University Press.

Zhmud, Leonid (2006) *The Origin of the History of Science in Classical Antiquity*. Berlin: Walter de Gruyter.

Index

Page numbers in *italics* refer to illustrations.

actors' categories, 94n33
Aetna (poem), 2, *30*, 31-33, 45-55, 80: author, identity of, 31, 32, 33, 45, 81; borrowings in, 54; corruption of text, 103n54; editions/translations of, 99n7, 102n48, 102n51; moral message of, 55; mythology in, 33, 35, 46, 54, 55; possibly requested by a patron, 45; purpose of, 32, 45; structure of, 51-52; subject matter of, 45-47, 48-51, 53, 64-65, 83, 85, 103n60
agriculture/gardening, 31, 45, 52, 93n22, 105n7
akousmata (things heard), 14-15. See also oral communication
Alexander of Aphrodisias, 29
Alexandria, 28
allegory, in poetry, 34, 42, 99n11
analogies, use of, 51
Anaximander, 14, 89
Anaximenes, 14, 89
Andronicus of Rhodes, 95n58, 96n66
Antigonus Gonatas of Macedonia, 40
Aphrodite (goddess), 101n37
Apollo (god), 44-45, 55
aporetic style, 22. See also problem texts
Aratus, 40, 45: *Phaenomena*, 40, 52; translations of works by, 109n6
Archimedes, 25, 98n3, 95n51: his bath, 95n51
Archytas of Tarentum, 14
Aristarchus: *On the Sizes and Distances*, 64
Aristophanes: *The Clouds*, 94n35
Aristotle, 18, 20, 91n6, 95n58, 96n62: Jonathan Barnes on, 18, 20; commentaries on, 21, 28-29, 100n15, 111n13; death of, 60; dialogues of, 60, 104n7, 107n39; on Empedocles, 38-39; *Eudemian Ethics*, 18, 95n53, 96n61; on knowledge (*epistēmē*), 7; lectures of, 18; library of, 15-16, 17; *Metaphysics*, 18, 22-23, 34, 92n21, 99n9; *Meteorology*, 24; method of study, 16, 17, 28, 51; *On the Generation of Animals*, 38; *On the Heavens*, 29, 100n15; on mythology, 33-34; *On the Pythagoreans*, 15; *Physics*, 18; Simplicius' commentary on, 28-29, 100n15, 111n13; *Poetics*, 38, 59-60, 99n8; *Posterior Analytics*, 58, 97n78; *Prior Analytics*, 19; *Problems*, 23; Pseudo-Aristotle: *Problems* (*Problemata*), 23, 25, 97n75; *Topics*, 17
Asclepius of Tralles, 27-28
Asmis, E.: *Epicurus' Scientific Method*, 103n60
astronomical instruments, 73, 107n48
astronomy, 11, 27, 35-36, 43, 47, 89: cosmology, 57, 72, 73, 99n12; moon, 66-67, 75-76, 105n1; planets, 35, 107n48; in Plato: *Timaeus*, 71, 72-73; in poetry, 43-45, 48, 52. See also Plutarch: *On the Face on the Moon*
Athens, 15, 70: library in, 87, 110n4
audiences, 15, 21-22: importance of, 13-14. See also oral communication
authors, 79, 81: the "ancients," 83, 85, 88, 109n14; as conveyors of knowledge, 108n5; precedents, reference to, 82-83, 85, 89; readers and, 80, 81-82, 83, 88, 108n5. See also books/texts; genres, choice of; individual authors; readers
Avienus, 109n6

Bacchus (god), *frontis*
Barnes, Jonathan, 18, 20
bath. See Archimedes
biographies, 11, 16, 108n4
Boethius, 27
books/texts, 87-90: alterations/corrections to, 89; codices, 88-89, 108n5; dedicatees, 80; ease of use of, 89; editors of, 20, 85, 96n66; fiction, 3; intertextuality, 82-83;

| 131

papyrus rolls, 87, 88, 89, 110nn2-3; parchment, 88; publication process, 88; readers' impact on, 80, 81; structure of, 87, 88; survival of, 89-90; translations of, 80, 85, 90, 109n6, 109n9. *See also* authors; genres; individual authors of; libraries; readers

Buxton, Richard: *From Myth to Reason?...*, 92n7, 92n18

Calame, Claude, 3-4, 92n18
calendars, 52
Carthage, 62, 69, 107n37
Cato the Elder: *On Agriculture*, 93n22
catoptrics. *See* mirrors
Cavallo, Guglielmo, 108n5
chanting. *See* incantations
Cherniss, Harold, 63, 67-68, 74, 105n17, 106n23, 106n36
Chrysippus (of Soli), 102n43
Cicero, Marcus Tullius, 7, 97n75, 109n6: dialogues, 60-61, 105n7, 107n39; *Letters to Atticus*, 58; *On the Nature of the Gods*, 102n43; *Tusculan Disputations*, 103n56
Clay, Diskin: *Lucretius and Epicurus*, 101n37
Cleanthes (of Assos), 102n43
Clearchus, 64
Cleomedes, 19: *The Heavens*, 19, 21
codices (the codex), 88-89, 108n5
Columella: *On Agriculture*, 31, 45
commentaries, 11, 21-22, 28-29, 80, 82, 85, 89: on Aristotle, 21, 28-29, 100n15, 111n13; on Homer, 17, 22, 28; on mathematics, 28
communication. *See* oral communication; scientific communication; written communication
communities, intellectual/scientific, 81-82, 109n9: definition of, 83, 84; disagreement within, 84; membership of, 83-84, 85. *See also* philosophical schools

Conte, Gian Biagio, 8-9: *Genres and Readers*, 8
conversation, 82: dialogues as, 17, 58, 59-60, 77, 82, 100n23; dinner-table topics, *see* philosophy
Copernicus, Nicolaus, 57
Cornford, F. M., 73
correspondence. *See* letters/correspondence
cosmology, 57, 72, 73, 99n12: Copernican, 57; Plutarch on, *see* Plutarch: *On the Face on the Moon*. *See also* astronomy
creation myths, 43-44, 71, 72-73, 75-76
Cronus (god), 62, 66, 67

Darwin, Erasmus: *The Botanic Garden* (poem), 104n1
death, mythic account of, 66-67, 75
Delphic shrine, 61
Democritus, 16
demonstrations, 94n33
Depew, Mary, and Dirk Obbink (eds.): *Matrices of Genre*, 93n25
Descartes, René, 36
dialogue(s) (*dialogos*), 2, 11, 12, 17, 19, 57-78: author's choice of, as a genre, 58-59, 60, 76-78, 80; of Aristotle, 60, 104n7, 107n39; author's voice in, 59, 61; of Cicero, 60-61, 105n7, 107n39; conversation as, 17, 58, 59-60, 77, 82, 100n23; definition of, 58, 76; disagreements in, 60, 61; of Galileo, 77-78; influence/importance of, 76-77, 78; monologues/speeches in, 60, 61, 107n39; myth in, 57, 58, 61, 62, 63, 66-68, 69, 71; of Plato, 17, 57, 59, 60, 70, 71-72, 76-77, 104n7, 105n18, 108n5 (*see also* Plato: *Timaeus*); of Plutarch, 60, 61, 104n7, 106n35 (*see also* Plutarch: *On the Face on the Moon*); problem texts as, 20; purpose of, 60-61, 76-78; stranger's role in, 62, 66-68, 105n18, 106n30; while walking, 65-66

Diodorus Siculus, 96n60
Diogenes, Antonius: *The Wonders Beyond Thule*, 105n16
Diogenes Laertius, 15, 16, 102n43, 111n13: "Life of Democritus," 16; "Life of Epicurus," 25; his *symbola*, 94n42
Diomedes: *Ars grammatica*, 101n30
Dionysius of Halicarnassus, 96n60
Dirlmeier, Franz, 19
discovery/invention (*heurēma*), 16
innovation, 84-85, 109n12
divinity/divinities, 35, 43-44
doxography, 16, 17
drama, 9. *See also* oral communication; poetry
Duff, David: *Modern Genre Theory*, 8

earth, as a living body, 49-50
earthquakes, 49
editors, 20, 85, 96n66. *See also* books/texts
education. *See* study; teachers/teaching
eisagōgai (introductory texts), 27-28: elements, 101n37
Empedocles, 4, 6, 28, 38-39, 99n8: Aristotle on, 38-39; Lucretius: *De rerum natura* influenced by, 101n93; *On nature* (poem), 101n93
encyclopaedias, 11
Epicureans, 10, 35, 93n29: beliefs of, 42-43, 48, 49; on ethics, 43; on gods 42-4; Lucretius as, 41-42, 43, 48: on natural philosophy, 48, 50-51
Epicurus, 93n29, 103n60: Diogenes Laertius: "Life of ...", 25; letters of, 25, 26-27, 103n59
Eratosthenes of Cyrene, 25, 39, 97n84, 98n86
Etna, *frontis.*, 45, 46-47, 48-49, 51. *See also* Aetna
Euclid: *Elements*, 25
Eudoxus, 40, 107n48
Euripides, 95n45

fiction, 3: genre fiction, 93n23. *See also* books/texts; stories

filial piety, 54-55, 103n69
Fleet, Barrie, 98n92
Frasca-Spada, Marina, and Nick Jardine: *Books and the Sciences in History*, 109n7
Furley, David, 101n37

Gadamer, Hans Georg, 76-77: *Truth and Methods*, 77
Galen (of Pergamum), 88
Galileo Galilei: dialogues of, 77-78; Johannes Kepler, correspondence with, 85; *Two Chief World Systems*, 57; *Two New Sciences*, 57
Gaye, R. K., 98n92
Geminus: *Introduction to the Phenomena*, 27
genres, 1-2, 6, 8-29: author's choice of, 31, 37-38, 39, 45-46, 52, 53,55, 60, 76, 77-78, 79, 80, 81, 82-83, 86; definition of, 8, 76, 93n23; differences between, 12; form/content, 9-11, 76; literary, 8, 9, 82, 84, 89, 93n23; oral, *see* oral communication; types of, 11-29, 58-59, 76-78, 108n4; written, *see* written communication. *See also* Plutarch: *On the Face on the Moon*
Germanicus, 109n6
gods. *See* Epicureans on; individual gods/goddesses
Goodyear, F. R. D., 53, 102n48
Goold, G. P., 102n44
Grafton, Anthony, 93n32, 96n70, 98n94, 104n2, 109n15
Greco-Roman science, 2, 7, 9, 88-90: development of, 5, 6; knowledge (*epistēmē*), 7, *see also* knowledge; George Sarton on, 3
Greek Anthology, 98n2
Greek miracle concept, 3, 5
Guthrie, W. K. C: *A History of Greek Philosophy*, 34

Hadrian, Emperor, 102n52
Hardie, Philip, 101n36, 104n66, 109n12
Hardie, R. P., 98n92

Harris, William V: *Ancient Literacy*, 95n45
Havelock, Eric A., 13
Heraclitus, 16
Herodotus: *History*, 12
Hesiod, 5, 6, 34-35, 89: dating of, 5, 92n16; importance/influence of, 31, 34, 35-36, 37, 40, 41; myth in poetry of, 34-35; *Theogony*, 40, 99n10; *Works and Days*, 40, 42, 89
heurematography, 16
Hine, Harry, 93n22, 102n49, 108n2: on volcanic activity, 48-49, 52
Hipparchus, 40, 69
Hippias of Elis, 16
Hippocrates, 21
historia, 94n33. *See also* written communication
Homer, 5, 6, 34-5, 38, 89, 99n8: *Homeric Hymns*, 41; *Iliad*, 3, 5, 34; importance/influence of, 31, 34, 35-36, 37, 41; mythology in, 34-35; *Odyssey*, 3, 5; oral nature 0; poetry of, 13-14; study of/commentaries on, 17, 22, 2
hymn/hymning (*hymnos*), 41. *See also* poetry
hypomnēma, 96n61. *See also* written communication

Iamblichus, 15, 27-28
ideas: communication of, 5, 21; study/collection of, 16, 17, 23, 97n74
incantations/chanting, 74, 107n50
innovation, 84-85, 109n12
invention. *See* discovery/invention; innovation

Jardine, Nick *see* Frasca-Spada, Marina, and Nick Jardine
Johansen, Thomas, 78
Johnson, William A., 82
journal articles, 1, 2, 10-11, 96n63. *See also* written communication

Keaney, John J. *see* Lamberton, Robert, and John J. Keaney
Kepler, Johannes, 66, 104n2, 109n9: Galileo, correspondence with, 85;

Plutarch, interest in, 85; *On the Face on the Moon*, Latin translation of, 57
Somnium (*Dream*), 57
knowledge (*epistēmē*), 7, 58, 68-70, 76, 92n20, 108n5: Aristotle on, 7; gaining of, 78; innovation as, 84-85; as interpretation, 76-78; types of, 7. *See also* study; teaching/teachers
Konstan, David, 99n10, 100n26

Lamberton, Robert, 60, 68, 76, 99n10, 106n36: and John J. Keaney: *Homer's Ancient Readers*, 99n10
language, Greek and Latin, 7
lecture notes, 18, 19-20, 21. *See also* written communication
lectures, 11, 12, 18, 79: as treatises, 19-20, 21. *See also* oral communication
letters/correspondence, 11, 25-27, 45, 47: circulation of, 25, 26; of Epicurus, 25, 26-27, 103n59; mathematical, 25, 108n4
Lewenstein, Bruce V., 91n1
libraries, 15-16, 17, 21-22: in Athens, 87, 110n4. *See also* books; readers
linguistic issues, 4, 6-7, 92n20
literacy levels, 15, 17, 18, 21
Lloyd, Geoffrey E. R., 18, 19, 59, 84, 91n5, 94n36, 94n39, 99n12,
logos. *See* reason
Longfellow, Henry Wadsworth: *Evangeline* (poem), 100n16
Lord, Albert, 13
Lowenstam, Steven, 91n3
Lucian, 104n7: *A True Story*, 105n16
Lucilius (friend of Seneca), 23, 45, 101n52
Lucretius, 10, 54, 103n60: as an Epicurean, 41-42, 43, 48
Lucretius: *De rerum natura* (*On the Nature of the Universe*) (poem), 2, 10, 31, 39, 40, 41-45, 79: allegory in, 42; influences on, 42-43, 101n27; myth in, 41-43; order of the lines, 101n36; proem, 41; subject matter of, 43, 50, 52, 55, 103nn59-60

134 | Index

Manilius, Marcus, 102nn44-45, 103n60
Astronomica, 40, 43-45, 103n60
mathēma, 92n21
mathematical letters, 25, 108n4
mathematics, 7, 9-10, 12, 17, 64-65, 92n21, 93n27, 103n56: arithmetic, 27-28; commentaries on, 28; geometry, 24, 25, 69, 103n56; introductory texts, 27-28; poetry on, 31; problem texts on, 24, 25
McLuhan, Marshall: *Understanding Media ...*, 93n31
medical texts, 92n19, 93n27, 94n34
medicine/cures, 74
Mercury (god), 44-45
meteorology, 23-24, 35-36, 43, 45, 47, 48, 49, 52, 89
meter. *See* poetic meter
mirrors (catoptrics), 64, 65, 106nn22-23
moon, 75-76, 105n16: possibility of life on, 66, 67; as resting place of the soul after death, 66-67, 75. *See also* astronomy; Plutarch: *On the Face on the Moon*
Morgan, Kathryn A: *Myth and Philosophy*, 92n18
Most, Glenn, 4-5, 8-9
Mourelatos, P. D: *The Route of Parmenides*, 37-38
myth (*mythos*), 3: of creation, 43-44, 71, 72-73, 75-76; of death, 66-68; definition of, 3-4, 6, 92n18; in dialogues, 57, 58, 61, 62, 63, 66-68, 69, 71; divinity/divinities in, 35, 43-44; Epicurean attitudes to, 42-43; importance of, 76; as lies, 32; of the moon, 66-67, 75; philosophy and, 3, 33-34, 35, 91n6; Plato on, 34, 58, 104n5; in poetry, 32, 33-36, 38, 41-46, 54, 55; reason (*logos*) and, 4; rejection of, 3, 4-5, 6, 57-58; science and, 2, 4, 6, 57-78; uses of, 35

natural philosophy, 7, 9, 12-13, 34, 35, 80, 92n20: analogy, use in, 51; Epicurean, 48, 50-51; Stoic, 102n43

natural world, poetry on, 31-32, 35-36, 40-44, 46-7, 98n2, 101n37. *See also* Aetna
Nicomachus of Gerasa, 28
Introduction to Arithmetic, 27-28
Nussbaum, Martha, 17, 20

Obbink, Dirk. *See* Depew, Mary, and Dirk Obbink
Oldroyd, D. R., 32
Olympiodorus, 29
oral communication: audience(s) for, 13-14, 15, 21-22; drama, 9; hearing/listening, 14; lectures, 11, 12, 79; notes on, 18, 19-20, 21; poetry, *see* poetry; press conferences, 1, 2, 29; things heard (*akousmata*), 14-15. *See also* written communication
Ovid, 45, 52-53: *Fasti*, 52

Paisley, P. B., 32
papyrus rolls, 87, 88, 89, 110nn2-3: Derveni papyrus, 91n5
Parmenides of Elea, 4, 6, 16, 36-38, 83: importance/influence of, 36
Parry, Milman, 13
patronage, 40, 45, 97n84, 98n86
Pherecydes of Syros, 4
Philolaus, 14
Philodemus, 93n29
Philoponus, John, 27-8, 29
philosophical schools, 23: Epicurean, *see* Epicureans; Peripatetic, 23, 69, 101n30; Pythagorean, 14-15; Stoic, 43-4, 64, 68, 69, 91n6, 102n43, 103n60. *See also* communities; study
Philosophical Transactions of the Royal Society, 11
philosophy: communication of ideas in, 5, 21; myth and, 3, 33-34, 35, 91n6; natural, 7, 9, 12-13, 35, 92n20; poetry and, 6, 21, 36-45; reason (*logos*) and, 3, 91n6; topic of conversation, suitability as a, 100n23
Phokos of Samos, 111n13
physis; *physikoi*, 7, 12

Pindar: *Pythian Ode*, 52
Pisistratus, 15
planets, 35, 107n48. *See also* astronomy
Plato, 4, 19, 67, 70, 85, 91n5: *Cratylus*, 34; dialogues of, 17, 57, 59, 60, 70, 71-72, 76-77, 104n7,105n18, 108n5 (*see also* Plato: *Timaeus*); educational system of, 105n10; Hans Georg Gadamer on, 76-77; *Gorgias*, 96n61; *Meno*, 77; on myth 34, 58, 104n5; *Phaedo*, 107nn50-51; *Phaedrus*, 108n5; *Protagoras*, 71-72, 92n12; *Republic*, 91n6, 104n5, 105n10, 107n50; subject matter of, 57, 59, 105n10; *Theaetetus*, 16, 34, 96n61; *Timaeus-Critias*, 70, 75, 107n39
Plato—*Timaeus*, 2, 57, 58, 70-76, 92n8: creation myth in, 71, 72-73,75-76; dating of, 70; as a dialogue, 12, 58, 70, 75, 78; incomplete, 74, 106n23; as a monologue, 60, 70; mythology in, 58, 61; Plutarch's interest in, 61; Simplicius on, 29; Socrates represented in, 70, 71; structure of, 60, 70-71, 72 (*see also* as a dialogue *above*); subject matter of, 61, 70, 71, 72, 74-76
Plato's Academy, 23;
Platthy, Jeno: *Sources on the Earliest Greek Libraries ...*, 94n45
Pliny the Elder, 36: *Natural History*, 87
Pliny the Younger, 88
Plotinus, Porphyry: *Life of ...*, 21-22, 29, 109n15
Plutarch, 38, 58, 81, 97n75, 104n2: as author, 82-83; dialogues of, 60, 61, 104n7, 106n35 (*see also* Plutarch—*On the Face on the Moon*); *How a Youth Should Listen to Poems*, 81-82; Johannes Kepler's interest in, 85; life of, 61; on mythology, 78; *On the Festival of Images ...*, 78; *On Listening to Lectures* (*De audiendo*), 73,106n23; as Platonist, 61; reading habits of, 109n8
Plutarch—*On the Face on the Moon*, 2, 58, 61-63,67-69, 80, 98n4: ambiguities in, 68; author's voice in, 63, 106n35; corruption of text, 61; as a dialogue, 58, 61-62, 75, 76, 77-78; as a dialogue within a dialogue, 61, 62-63, 75, 78; knowledge displayed in, 68-70; manuscripts of, 61, 105n17; myth in, 58, 61, 62, 53, 66-8, 69, 73-74,75-76, 78, 107n51; narrator in, 61-62, 63, 67-69; stranger, role of in, 62, 66-68, 105n18, 106n30; structure of, 61-63, 68, 105n17, 106n36 (*see also* as a dialogue *above*); subject matter of, 61, 62, 63, 64-68, 73-74, 75-76, 83
Pluto (god), 54
poetic meter, 36, 41, 100n16
poetic muses, 55
poetry, 2, 3, 5, 9, 17, 31-55: allegory in, 34, 42, 99n11; astronomy in, 43-45, 48, 52 ; choice of, as a genre, 31, 37-38, 39, 45-46, 52, 53,55, 80; criticism of, 33, 34, 53-54; dating of, 5; didactic, 21, 40, 41, 52, 53-54, 101n30, 109n12; hymns/hymning, 41; importance of, 31, 39; interpretation of, 34; mathematics in, 31; mythology in, 32, 33-36, 38, 41-46, 54, 55 (*see also Aetna*); on natural world, 31-32, 35-36, 40-44, 46-7, 98n2,101n37; as performance, 17; philosophy and, 6, 21, 36-45; proem (beginning a poem), 37, 41; science and, 2, 5-6, 10, 11, 31-55, 58; teaching texts as, 12; as truth, 33; uses of, 20-21, 39-40. *See also* individual poets/poems
Polybius, 96n60
Porphyry: *Life of Plotinus*, 21-22, 29, 109n15
Posidonius, 65
pragmateiai, 19, 95n60, 96n67. *See also* written communication

problem texts, 11, 22-25: definition of, 22; as dialogues, 20; format of, 23, 25; on mathematics, 24, 25
zētēmata, 97n76
Pseudo-Aristotle. *See* Aristotle
Ptolemy: *Almagest*, 109n14
Ptolemy, King of Egypt, 97n84, 98n86
Pythagoras, 14, 27: Johannes Kepler's interest in, 85; maxims, 15
Pythagoreans, 14-15

question-and-answer texts. *See* problem texts

readers, 79, 80, 85, 108n5, 109n9: authors and, 80, 81-82, 83, 108n5; as a group, 81-82, 109nn8-9; impact of on texts, 80, 81-82; of problem texts, 96n62; types of, 81. *See also* books/texts; libraries
reading aloud, 81-82, 109nn8-9. *See also* oral communication
reason (*logos*), 58-59: definition of, 3, 4; as discourse, 4; myth and, 4, 58; philosophy and, 3; types of, 3
religion, 35: Christianity, influence of, 88, 110n8
Reynolds, L. D., and N. G. Wilson: *Scribes and Scholars*, 110nn5-6
rhetoric, 9
Roberts, Colin H., and T. C. Skeat: *The Birth of the Codex*, 88, 110n8
Ross, W. D., 99n9
Royal Society. *See Philosophical Transactions ...*

Sarton, George, 3, 5
Schiesaro, Alessandro, 101n30
Schofield, Malcolm, 36, 38
science, 9-10: definition of, 7, 12-13, 57-58, 92n20; history of, 9-10, 11, 32; myth and, 2, 4, 6, 57-78. *See also* Greco-Roman science; mathematics
scientia, 92n20
scientific communication, 1-29: genres/forms of, 1-2, 6, 9-29, 79-90 (*see also* individual genres); in Greco-Roman world, 2, 5, 82-83. *See also* oral communication; written communication
scientific practice, 32, 47, 48, 49, 50-51, 103n56: observation, 48, 49, 50-51
Scriptores Historiae Augustae, 102n52
Sedley, David, 103n60, 107n48: *Creationism and its Critics ...*, 71-72; *Lucretius and the Transformation of Greek Wisdom*, 43, 101n37
Seneca, 25, 50, 52-53, 102n52: *Epistles*, 102n52; *Natural Questions (Questions about Nature)*, 23, 24-25, 55, 102n50
Severus, Cornelius, 45, 53
Sextus Empiricus, 99n15
Sider, David, 96n61
Simplicius, 77, 99n15: commentary on Aristotle, *Physics*, 28-29, 100n15, 111n13; Parmenides and, 83
Skeat, T. C., *see* Roberts, Colin, and T. C. Skeat
skholai (lectures), 19
Sluiter, Ineke, 98n93
Smith, Robin, 95n53
Snyder, Gregory, 15
Socrates, 59: Hans Georg Gadamer on, 76-77; as a voice in Plato's dialogues, 59, 60, 70, 71
Solmsen, Friedrich, 54
Sophocles, 82, 106n26
soul, 67, 75
Stephens, Susan A., 110n2
Stoicorum Veterum Fragmenta, 103n60
Stoics, 43-44, 64, 68, 69, 91n6, 103n60: on natural philosophy, 102n43
stories, 3-4. *See also* myth
Strabo, 46: *The Geography*, 39
stranger, role of in Greek texts, 62, 66-68, 105n18, 106n30
students/followers, 14. *See also* teaching/teachers
study: methods of, 16-17; of natural world, importance of, 47-48,

Index | 137

103n56. *See also* philosophical schools
Sudhaus, Siegfried, 54
suggrama, 19, 96n61. *See also* written communication

Taylor, A. E., 72
teaching/teachers, 29, 80, 98n95: of ideas, 16, 17, 23, 97n74; Plato's interest in, 105n10; poetry used in, 39. *See also* philosophical schools
teaching texts, 11, 27-28, 57: poems as, 12; systematic, 96n61
texts. *See* books/texts
Thales of Miletus, 14, 89, 111n13
Thalman, William G: *Conventions of Form ... in Early Greek Epic Poetry*, 101n32
Theagenes of Rhegium, 34
Theophrastus, 49, 50, 103n59
Todorov, Tzvetan, 8
Toohey, P: *Epic Lessons ...*, 102n45
topos (literary commonplace), 52-53
Tractatus Coislinianus, 101n30
translations, 80, 85, 90, 109n6, 109n9. *See also* books/texts
treatises, 11, 18-22, 94n33: definition of, 20; lectures as, 19-20, 21
truth, concept of, 77, 78
Turner, Eric G: *The Typology of the Early Codex*, 110n10

understanding. *See* knowledge

van der Eijk, Philip, 20, 92n20, 93n27, 94n39, 97n79, 104n3
Varro: *De re rustica*, 105n7
Venus (goddess), 41-43, 55
Verginius Rufus, 88
Vesuvius. *See Aetna*; Etna
Virgil, 45, 52-53, 54: *Georgics*, 40, 52
Vocanius Romanus, 110n7
volcanic activity, Harry Hine on, 48-49, 52. *See also Aetna*; Etna
Volk, Katherina: *The Poetics of Latin Didactic ...*, 101n30
von Staden, Heinrich, 91n5, 98n93

West, M. L., 92n16
Willmoth, Frances, 93n30
Wilson, N. G. *See* Reynolds, L. D., and N. G. Wilson
wind(s), 49, 103n59
Worton, Michael, 109n11
written communication, 1, 2, 15-29, 79: commentaries, 11, 21-22, 28-29, 80, 82, 85, 89; *historia*, 94n33; *hypomnēma*, 96n61; journal articles, 1, 2, 10-11, 96n63; lecture notes, 18, 19-20, 21; letters, 11, 25-7, 45, 77, 108n4; *pragmateiai*, 19, 95n60, 96n67; problem texts, 11, 22-25, 96n62; 97n76; *suggramma*, 19, 96n61; teaching texts, 11, 12, 27-28, 57, 96n61; texts read to an audience, 15, 94n33; treatises, 11, 18-22, 94n33. *See also* dialogues, oral communication, poetry

Xenophon, 59

Zeno of Citium, 102n43
zētēmata, 97n76. *See also* problem texts